Concise Introduction to
LINEAR ALGEBRA

Concise Introduction to
LINEAR ALGEBRA

Qingwen Hu

CRC Press
Taylor & Francis Group
Boca Raton London New York

CRC Press is an imprint of the
Taylor & Francis Group, an **informa** business
A CHAPMAN & HALL BOOK

CRC Press
Taylor & Francis Group
6000 Broken Sound Parkway NW, Suite 300
Boca Raton, FL 33487-2742

First issued in paperback 2020

Version Date: 20170822

ISBN 13: 978-0-367-65770-3 (pbk)
ISBN 13: 978-1-138-04449-4 (hbk)

Visit the Taylor & Francis Web site at
http://www.taylorandfrancis.com

and the CRC Press Web site at
http://www.crcpress.com

To my students.

Contents

Preface

This book provides sufficient materials for a one-semester linear algebra course at the sophomore level. It is based on the lecture notes for the linear algebra course that the author taught several years to undergraduate students in science and mathematics at the University of Texas at Dallas. The level and pace of the course can be adjusted by balancing the time for theoretical illustration and that for computational aspects of the subject. The author usually taught up to Chapter 7, spending one lecture per section on average, while the remaining two chapters can be left for students' reading homework or supervised individual study.

It seems that many undergraduate students have only one linear algebra course before graduation, and may have missed many important topics of linear algebra which may be remedied later by self-studying on demand. This book is written to accommodate the needs for classroom teaching in order to effectively deliver the essential topics of the subject, and for self-studying beyond a first linear algebra course.

The following is an introduction to each chapter of the book.

1. Chapter 1 deals with vectors, linear combinations and dot products in $\mathbb{R}^n$. In Section 1.3 we discuss matrix representations for linear systems and for elementary row operations.

2. Chapter 2 illustrates Guassian elimination and Gauss–Jordan elimination for solving linear systems, along with basic matrix theory, LU-decomposition and permutation matrices.

3. Chapter 3 starts with four subspaces of $\mathbb{R}^n$ associated with a real matrix. Then we discuss bases and dimensions of general vector spaces.

4. Chapter 4 deals with orthogonality between subspaces. Related topics include matrix representation of orthogonal projection, least squares solutions, Gram–Schmidt process and QR-decomposition.

5. Chapter 5 presents an axiomatic method of determinants which naturally leads to the permutation formula, co-factor expansion, product formula and Cramer's rule.

6. In Chapter 6 we introduce the notions of eigenvalues and eigenvectors which open the door for more applications of linear algebra, including the immediate application on diagonalizability, spectral decomposition of symmetric real matrices, quadratic forms, positive definite matrices and Rayleigh quotient.

7. Chapter 7 continues to discuss the application of eigenvalues and eigen-vectors and presents singular value decomposition of general matrices. Principal component analysis is also introduced as a real-world application of linear algebra.

8. Chapter 8 discusses the matrix representation, range and null spaces for linear transformations on general vector spaces. Then we introduce invariant subspaces, decomposition of vector spaces and Jordan normal form and its computation, where the treatment of the Jordan normal form does not require a formal exposition of polynomial theory.

9. Chapter 9 presents basic theory of linear programming along with the simplex method which is another concrete real-world application of linear algebra and which has been widely used in management and industry.

The book contains typical topics for linear algebra courses and can be used in many ways depending on the different mathematical background of the audiences. The book provides limited examples and exercises, while it is best used for readers who would like to have a broad coverage of the topics of linear algebra and who are motivated to customize questions for the materials of each section. Comments and suggestions from readers are highly appreciated and are welcome to be sent by e-mail to `qingwen@utdallas.edu`.

Qingwen Hu
January 2017

Chapter 1

Vectors and linear systems

A central goal of linear algebra is to solve systems of linear equations. We have seen the simplest linear equation $ax = b$, where $x \in \mathbb{R}$ (the symbol "$\in$" means "in") is the unknown variable and a, $b \in \mathbb{R}$ are constants. It is known that there are three scenarios for the solutions: 1) if $a \neq 0$, there is a unique solution $x = \frac{b}{a}$; 2) if $a = 0$, $b \neq 0$, there is no solution; 3) if $a = b = 0$, there are infinitely many solutions. We are then motivated to investigate systems of equations with multiple unknown variables. The following system

$$\begin{cases} x + 2y + 3z = 3 \\ 2x + 5y + 8z = 9 \\ 3x + 6y + 18z = 18 \end{cases} \qquad (1.1)$$

is a system of linear equations with three equations and three unknowns. In this chapter, we learn how to use vectors to represent a linear system and learn the ideas of elimination which will be applied to solve systems of linear equations. The general form of linear systems is as follows:

$$\begin{cases} a_{11}x_1 + a_{12}x_2 + \cdots + a_{1n}x_n = b_1, \\ a_{21}x_1 + a_{22}x_2 + \cdots + a_{2n}x_n = b_2, \\ \qquad \cdots\cdots\cdots\cdots\cdots\cdots\cdots\cdots\cdots \\ a_{m1}x_1 + a_{m2}x_2 + \cdots + a_{mn}x_n = b_m, \end{cases} \qquad (1.2)$$

where $x = (x_1, x_2, \cdots, x_n) \in \mathbb{R}^n$ is the unknown vector in n-dimensional Euclidean space; $a_{i,j}$ and b_i with $i \in \{1, 2, \cdots, m\}$, $j \in \{1, 2, \cdots, n\}$ are constants.

1.1   Vectors and linear combinations

Before we discuss how to solve general linear systems, we use system (1.1) as a prototype to introduce the machinery of vectors. One may rewrite system (1.1) as

$$x \begin{bmatrix} 1 \\ 2 \\ 3 \end{bmatrix} + y \begin{bmatrix} 2 \\ 5 \\ 6 \end{bmatrix} + z \begin{bmatrix} 3 \\ 8 \\ 18 \end{bmatrix} = \begin{bmatrix} 3 \\ 9 \\ 18 \end{bmatrix}. \tag{1.3}$$

System (1.3) makes sense only if we have defined addition and scalar multiplication of vectors in Euclidean spaces, where we have identified the vector (x_1, x_2, x_3) with the column of numbers

$$\begin{bmatrix} x_1 \\ x_2 \\ x_3 \end{bmatrix},$$

which is called a column **matrix.** In what follows we will always regard a vector in $\mathbb{R}^n$ as a column matrix.

Definition 1.1.1. Let $x = (x_1, x_2, \cdots, x_n)$, $y = (y_1, y_2, \cdots, y_n)$ be vectors in $\mathbb{R}^n$, α a scalar. We define addition $x + y$ and scalar multiplication αx by

$$x + y = (x_1 + y_1, x_2 + y_2, \cdots, x_n + y_n),$$
$$\alpha x = (\alpha x_1, \alpha x_2, \cdots, \alpha x_n).$$

Definition 1.1.2. Let $x_1, x_2, \cdots, x_n \in \mathbb{R}^N$ be vectors, and $c_1, c_2 \cdots, c_n \in \mathbb{R}$ be scalars. We call

$$c_1 x_1 + c_2 x_2 + \cdots c_n x_n$$

a **linear combination** of $x_1, x_2, \cdots, x_n$.

System (1.3) now can be interpreted as finding a proper linear combination of the vectors $(1, 2, 3)$, $(2, 5, 6)$ and $(3, 8, 18)$ to produce the given vector $(3, 9, 18)$ on the right hand side. Certainly we can also interpret it as finding the common point (x, y, z) of three planes determined by each of the equations. If we visualize a linear system with this interpretation of a linear system, we obtain a row picture, while with the previous one, a column picture.

Example 1.1.3.   1. Let $v = \begin{bmatrix} 1 \\ 1 \end{bmatrix}$, $w = \begin{bmatrix} 1 \\ 3 \end{bmatrix}$. Then

$$3v + 5w = 3 \begin{bmatrix} 1 \\ 1 \end{bmatrix} + 5 \begin{bmatrix} 1 \\ 3 \end{bmatrix} = \begin{bmatrix} 8 \\ 18 \end{bmatrix}$$

is a linear combination of v and w.

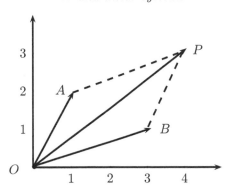

FIGURE 1.1: Slope of $\overrightarrow{OA} = \frac{2-0}{1-0} = 2$, $\overrightarrow{BP} = \frac{4-2}{2-1} = 2$. Slope of $\overrightarrow{OB} = \frac{3-0}{1-0} = 3$, $\overrightarrow{AP} = \frac{4-1}{3-2} = 3$.

2. Let $v = \begin{bmatrix} 1 \\ 2 \end{bmatrix}$. Then $\{cv : 0 \leq c \leq 2\}$ represents a line segment from $(0, 0)$ to $(2, 4)$ in $\mathbb{R}^2$.

3. Let $v = \begin{bmatrix} 1 \\ 0 \end{bmatrix}$ and $w = \begin{bmatrix} 0 \\ 1 \end{bmatrix}$. Then $\{cv + dw : c \in \mathbb{R}, d \in \mathbb{R}\}$ represents the whole two dimensional plane $\mathbb{R}^2$.

4. Let $v = \begin{bmatrix} 1 \\ 1 \\ 0 \end{bmatrix}$ and $w = \begin{bmatrix} 1 \\ 1 \\ 1 \end{bmatrix}$. Then $S = \{cv + dw : c \in \mathbb{R}, d \in \mathbb{R}\}$ represents a two dimensional plane in $\mathbb{R}^3$, but not the whole space $\mathbb{R}^3$, because there exists the vector $\begin{bmatrix} 1 \\ 2 \\ 3 \end{bmatrix}$ which is not in S.

□

Example 1.1.4. (The parallelogram law for vector addition) A vector $x = (x_1, x_2, \cdots, x_n) \in \mathbb{R}^n$ can be visualized by the directed line segment $\overrightarrow{OA}$ from the origin $O = (0, 0, \cdots, 0)$ to the point $A = (x_1, x_2, \cdots, x_n) \in \mathbb{R}^n$. If we denote the end point of the vector $y = (y_1, y_2, \cdots, y_n)$ by B and that of $x + y$ by P, then we have a parallelogram $OAPB$, with OA parallel to BP and AP parallel to OB, since the opposite segments have the same slopes.

□

Exercise 1.1.5.

1. Let $u = \begin{bmatrix} 1 \\ 1 \end{bmatrix}$, $v = \begin{bmatrix} 2 \\ 3 \end{bmatrix}$. i) Sketch the directed line segments in $\mathbb{R}^2$ that

represents u and v, respectively; ii) Use the parallelogram law to visualize the vector addition $u + v$; iii) Find $2u$, $2u + 5v$ and $2v - 5u$; iv) Solve the system of equations $xu + yv = \begin{bmatrix} -1 \\ 1 \end{bmatrix}$ for $(x, y) \in \mathbb{R}^2$ and draw the row picture and the column picture.

2. Let $u = \begin{bmatrix} 1 \\ 1 \end{bmatrix}$, $v = \begin{bmatrix} 2 \\ 3 \end{bmatrix}$. Is $w = \begin{bmatrix} 1 \\ -1 \end{bmatrix}$ a linear combination of u and v?

3. Is it true every vector $(x, y) \in \mathbb{R}^2$ can be represented as a linear combination of $v = (1, 0)$ and $w = (1, 1)$?

4. Find vectors $u, v, w \in \mathbb{R}^3$ such that the following system

$$\begin{cases} x + z = 1 \\ 2x + 5y + 8z = -1 \\ x + y = 1 \end{cases}$$

can be rewritten as $xu + yv + zw = b$, where $b = (1, -1, 1)$.

5. Show that $\mathbb{R}^2 = \left\{ x \begin{bmatrix} 1 \\ 0 \end{bmatrix} + y \begin{bmatrix} 0 \\ 1 \end{bmatrix} : x \in \mathbb{R}, y \in \mathbb{R} \right\}$.

1.2   Length, angle and dot products

In order to discuss geometry in Euclidean spaces, we introduce the notions of length and angle, which can be defined with dot products.

Definition 1.2.1. Let $x = (x_1, x_2, \cdots, x_n)$, $y = (y_1, y_2, \cdots, y_n)$ be vectors in $\mathbb{R}^n$; the dot product $x \cdot y$ is defined by

$$x \cdot y = x_1 y_1 + x_2 y_2 + \cdots + x_n y_n = \sum_{i=1}^{n} x_i y_i.$$

Example 1.2.2.   1. Let $v = \begin{bmatrix} 1 \\ 1 \end{bmatrix}$, $w = \begin{bmatrix} 2 \\ 3 \end{bmatrix}$. Then

$$v \cdot w = \begin{bmatrix} 1 \\ 1 \end{bmatrix} \cdot \begin{bmatrix} 2 \\ 3 \end{bmatrix} = 1 \cdot 2 + 1 \cdot 3 = 5.$$

2. Let $v = \begin{bmatrix} 1 \\ 1 \\ 0 \end{bmatrix}$ and $w = \begin{bmatrix} 1 \\ 1 \\ 1 \end{bmatrix}$. Then $v \cdot w = 1 \cdot 1 + 1 \cdot 1 + 1 \cdot 0 = 2$.

3. Let $v = \begin{bmatrix} 1 \\ 1 \end{bmatrix}$, $w = \begin{bmatrix} 1 \\ -1 \end{bmatrix}$. Then

$$v \cdot w = 1 \cdot 1 + 1 \cdot (-1) = 0.$$

We say v and w are orthogonal to each other and write $v \perp w$.

4. Consider the distance from $A = (1, 2)$ to the origin O. We have

$$\|\overrightarrow{OA}\| = \sqrt{(1 - 0)^2 + (2 - 0)^2}$$
$$= \sqrt{1 \cdot 1 + 2 \cdot 2}.$$

If we denote by v the vector $\overrightarrow{OA}$, we have the length of v

$$\|v\| = \sqrt{v \cdot v}.$$

5. Consider unit vectors $u, v \in \mathbb{R}^2$. Then there exist $\alpha, \beta \in [0, 2\pi)$ such that

$$u = (\cos\alpha, \sin\alpha), \quad v = (\cos\beta, \sin\beta).$$

Then we have

$$u \cdot v = \cos\alpha\cos\beta + \sin\alpha\sin\beta = \cos(\alpha - \beta).$$

$\square$

One can check directly that dot product satisfies the following

Lemma 1.2.3. Let $u, v, w \in \mathbb{R}^n$ be vectors. Then

$$u \cdot v = v \cdot u,$$
$$u \cdot (v + w) = u \cdot v + u \cdot w.$$

Definition 1.2.4. Let $v = (v_1, v_2, \cdots, v_n)$ be a vector in $\mathbb{R}^n$. The length $\|v\|$ of v is defined by

$$\|v\| = \sqrt{v \cdot v} = \left(\sum_{i=1}^n v_i^2 \right)^{\frac{1}{2}}.$$

A vector with unit length is called a unit vector.

Example 1.2.5.

Consider unit vectors $u, v \in \mathbb{R}^2$. Then there exist $\alpha, \beta \in [0, 2\pi)$ such that

$$u = (\cos\alpha, \sin\alpha), \quad v = (\cos\beta, \sin\beta).$$

Then we have

$$u \cdot v = \cos\alpha\cos\beta + \sin\alpha\sin\beta = \cos(\alpha - \beta).$$

There exists $\theta \in [0, \pi]$ such that $\cos\theta = \cos(\alpha - \beta)$. Then we call θ the angle between the vectors u and v.

Consider nonzero vectors u, $v \in \mathbb{R}^2$. Then $\frac{u}{\|u\|}$ and $\frac{v}{\|v\|}$ are unit vectors and there exist α, $\beta \in [0, 2\pi)$ such that

$$\frac{u}{\|u\|} = (\cos\alpha, \sin\alpha), \quad \frac{v}{\|v\|} = (\cos\beta, \sin\beta).$$

We have

$$\frac{u}{\|u\|} \cdot \frac{v}{\|v\|} = \cos(\alpha - \beta) = \cos\theta, \quad (1.4)$$

where $\theta \in [0, \pi]$ is the angle between $\frac{u}{\|u\|}$ and $\frac{v}{\|v\|}$. Notice that u and $\frac{u}{\|u\|}$ have the same direction, so do v and $\frac{v}{\|v\|}$. $\theta \in [0, \pi]$ is also the angle between u and v. By (1.4) we have

$$u \cdot v = \|u\| \|v\| \cos\theta,$$

where u and v can be zero. Then we have derived

Lemma 1.2.6. (Cosine formula) Let u, $v \in \mathbb{R}^2$. We have

$$u \cdot v = \|u\| \|v\| \cos\theta,$$

where $\theta \in [0, \pi]$ is the angle between u and v.

An immediate consequence of the cosine formula is that $|v \cdot w| = \|u\| \|v\| |\cos\theta| \leq \|u\| \|v\|$ which is the Schwarz inequality in $\mathbb{R}^2$. We show the general version of the Schwartz inequality in $\mathbb{R}^n$:

Lemma 1.2.7. (Schwarz inequality) Let u, $v \in \mathbb{R}^n$. We have

$$|u \cdot v| \leq \|u\| \|v\|.$$

Proof. The inequality is true if $v = 0$. We assume that $v \neq 0$ and let $w = u + tv$, $t \in \mathbb{R}$. Then $\|w\| \geq 0$ for every $t \in \mathbb{R}$. We have

$$0 \leq \|w\| = (u + tv) \cdot (u + tv)$$
$$= u \cdot u + 2(u \cdot v)t + (v \cdot v)t^2,$$

for every $t \in \mathbb{R}$. Therefore the discriminant of the quadratic polynomial $(u + tv, u + tv)$ of t satisfies

$$4(u \cdot v)^2 - 4(u \cdot u)(v \cdot v) \leq 0,$$

which is equivalent to $|u \cdot v| \leq \|u\| \|v\|$.   □

With the Schwarz inequality, we can then define angles between vectors in $\mathbb{R}^n$:

Definition 1.2.8. Let $u, v \in \mathbb{R}^n$. We define $\theta \in [0, \pi]$ such that

$$u \cdot v = \|u\|\|v\| \cos\theta,$$

the angle between u and v.

By properties of dot products and the Schwarz inequality, we have

> **Lemma 1.2.9. (Triangle inequality)** Let $u, v \in \mathbb{R}^n$. We have
>
> $$\|u + v\| \leq \|u\| + \|v\|.$$

Proof. We have

$$
\begin{aligned}
\|u + v\|^2 &= (u + v) \cdot (u + v) \\
&= u \cdot u + 2u \cdot v + v \cdot v \\
&\leq u \cdot u + 2\|u\| \cdot \|v\| + v \cdot v \\
&= \|u\|^2 + 2\|u\| \cdot \|v\| + \|v\|^2 \\
&= (\|u\| + \|v\|)^2.
\end{aligned}
$$

Therefore we have $\|u + v\| \leq \|u\| + \|v\|$. □

Exercise 1.2.10.

1. Let $u = \begin{bmatrix} 1 \\ 1 \end{bmatrix}$, $v = \begin{bmatrix} 2 \\ 3 \end{bmatrix}$. i) Find $u \cdot v$; ii) Find $\|u\|$ and $\|v\|$; iii) Find the angle θ between u and v; iv) Verify that $|u \cdot v| \leq \|u\|\|v\|$; v) Verify that $\|u + v\| \leq \|u\| + \|v\|$.

2. Find all possible real values of a such that the quadratic polynomial $x^2 + ax + 1$ has i) two positive roots; ii) two negative roots; iii) one negative and one positive root; iv) no real roots, respectively.

3. Let $u = \begin{bmatrix} 1 \\ 1 \end{bmatrix}$. Find all possible vectors w such that $u \perp w$, i.e., $u \cdot w = 0$.

4. Let $u = \begin{bmatrix} 1 \\ 1 \\ 1 \end{bmatrix}$, $v = \begin{bmatrix} -1 \\ 1 \\ 0 \end{bmatrix}$ and $w = \begin{bmatrix} 1 \\ 1 \\ -1 \end{bmatrix}$. i) Find $u \cdot v$ and $v \cdot w$. ii) Is it possible to find $(x, y) \neq (0, 0)$ such that $v = xu + yw$? Justify your answer.

5. Let $u, v \in \mathbb{R}^n$. Show that

$$|u \cdot v| = \|u\|\|v\|,$$

if and only if $v = 0$ or there exists a scalar $t \in \mathbb{R}$ such that $u = tv$.

1.3 Matrices

Recall that system (1.3) can be interpreted as finding a proper linear combination of the vectors $u = (1, 2, 3)$, $v = (2, 5, 8)$ and $w = (3, 6, 18)$ to produce the given vector $\mathbf{b} = (3, 9, 18)$ on the right hand side. That is, we are looking for scalars x, y, z such that

$$xu + yv + zw = \mathbf{b},$$

which looks to be a certain product between (u, v, w) and (x, y, z). To wit, we write

$$[u \quad v \quad w] \begin{bmatrix} x \\ y \\ z \end{bmatrix} = \mathbf{b},$$

which is a "row" multiplied by a "column." The reason why we put the letters for vectors horizontally becomes clear when we recover the values of u, v, w and $\mathbf{b}$:

$$\begin{bmatrix} 1 & 2 & 3 \\ 2 & 5 & 8 \\ 3 & 6 & 18 \end{bmatrix} \begin{bmatrix} x \\ y \\ z \end{bmatrix} = \begin{bmatrix} 3 \\ 9 \\ 18 \end{bmatrix}, \tag{1.5}$$

where we obtain a rectangular array of numbers called a **matrix**, and if u, v, w were placed vertically, we would not know how to place their values!

Let

$$A = \begin{bmatrix} 1 & 2 & 3 \\ 2 & 5 & 8 \\ 3 & 6 & 18 \end{bmatrix}, \ \mathbf{x} = \begin{bmatrix} x \\ y \\ z \end{bmatrix}, \ \mathbf{b} = \begin{bmatrix} 3 \\ 9 \\ 18 \end{bmatrix}.$$

System (1.3) becomes the familiar form of

$$A\mathbf{x} = \mathbf{b}. \tag{1.6}$$

By comparing system (1.3) with system (1.5), we know that the so far undefined product $A\mathbf{x}$ between matrices A and $\mathbf{x}$ essentially consists of rows of A taking dot products with $\mathbf{x}$. That is,

$$\begin{bmatrix} (\text{Row 1 of } A) \cdot \mathbf{x} \\ (\text{Row 2 of } A) \cdot \mathbf{x} \\ (\text{Row 3 of } A) \cdot \mathbf{x} \end{bmatrix} = \mathbf{b}.$$

Example 1.3.1.

$$\begin{bmatrix} 1 & 2 & 3 \\ -2 & -4 & 6 \end{bmatrix} \begin{bmatrix} 4 \\ -1 \\ 0 \end{bmatrix} = \begin{bmatrix} (1, 2, 3) \cdot (4, -1, 0) \\ (-2, -4, 6) \cdot (4, -1, 0) \end{bmatrix} = \begin{bmatrix} 2 \\ -4 \end{bmatrix}.$$

$$\begin{bmatrix} 3 & -2 \\ -1 & 0 \\ 2 & 5 \end{bmatrix} \begin{bmatrix} 2 \\ 1 \end{bmatrix} = \begin{bmatrix} (3, -2) \cdot (2, 1) \\ (-1, 0) \cdot (2, 1) \\ (2, 5) \cdot (2, 1) \end{bmatrix} = \begin{bmatrix} 4 \\ -2 \\ 9 \end{bmatrix}.$$

$$\begin{bmatrix} 1 & 0 & 0 \\ 0 & 1 & 0 \\ 0 & 0 & 1 \end{bmatrix} \begin{bmatrix} x_1 \\ x_2 \\ x_3 \end{bmatrix} = \begin{bmatrix} x_1 \\ x_2 \\ x_3 \end{bmatrix}.$$

□

Remark 1.3.2.

For an $m \times n$ matrix A, we write $A = (a_{ij})$ when we emphasize the general form of its entries. We also write $(A)_{ij}$, $A(i, j)$ or simply A_{ij} to denote the entry at the (i, j)-position.

If $A = (a_{ij})$ is an $n \times n$ square matrix, we call the entries a_{ii}, $i = 1, 2, \cdots, n$ the main diagonal entries. If every main diagonal entry of A is one, and every other entries are zero, that is,

$$a_{ij} = \begin{cases} 1 & \text{if } i = j \\ 0 & \text{if } i \neq j, \end{cases}$$

we call A an **identity matrix** and denote it by I. Note that

$$Ix = x \text{ for every } x \in \mathbb{R}^n.$$

Example 1.3.3. Let $u = (1, 0, 0)$, $v = (1, 1, 0)$ and $w = (1, 1, 1)$. $\mathbf{b} = (b_1, b_2, b_3)$. We solve system

$$A\mathbf{x} = \mathbf{b},$$

where

$$A = [u \quad v \quad w] = \begin{bmatrix} 1 & 1 & 1 \\ 0 & 1 & 1 \\ 0 & 0 & 1 \end{bmatrix}, \quad \mathbf{x} = \begin{bmatrix} x \\ y \\ z \end{bmatrix}.$$

That is, we solve

$$\begin{bmatrix} 1 & 1 & 1 \\ 0 & 1 & 1 \\ 0 & 0 & 1 \end{bmatrix} \begin{bmatrix} x \\ y \\ z \end{bmatrix} = \begin{bmatrix} b_1 \\ b_2 \\ b_3 \end{bmatrix}.$$

We notice that A is a triangular matrix in the sense that the nonzero entries are above the main diagonal. Such type of matrix is convenient for solving the system by **back substitution**. Namely, we first solve for z, then y and x. We obtain

$$\begin{bmatrix} x \\ y \\ z \end{bmatrix} = \begin{bmatrix} b_1 - b_2 - b_3 \\ b_2 - b_3 \\ b_3 \end{bmatrix}.$$

To have a solution resembling the solution $x = a^{-1}b$ of the single variable linear equation $ax = b$, $a \neq 0$, we wish to write (x, y, z) in terms of $\mathbf{b} = (b_1, b_2, b_3)$. We rewrite the solution as follows:

$$\begin{bmatrix} x \\ y \\ z \end{bmatrix} = \begin{bmatrix} b_1 - b_2 - b_3 \\ b_2 - b_3 \\ b_3 \end{bmatrix}$$

$$= \begin{bmatrix} b_1 \\ 0 \\ 0 \end{bmatrix} + \begin{bmatrix} -b_2 \\ b_2 \\ 0 \end{bmatrix} + \begin{bmatrix} -b_3 \\ -b_3 \\ b_3 \end{bmatrix}$$

$$= b_1 \begin{bmatrix} 1 \\ 0 \\ 0 \end{bmatrix} + b_2 \begin{bmatrix} -1 \\ 1 \\ 0 \end{bmatrix} + b_3 \begin{bmatrix} -1 \\ -1 \\ 1 \end{bmatrix}$$

$$= \begin{bmatrix} 1 & -1 & -1 \\ 0 & 1 & -1 \\ 0 & 0 & 1 \end{bmatrix} \begin{bmatrix} b_1 \\ b_2 \\ b_3 \end{bmatrix}.$$

Let

$$B = \begin{bmatrix} 1 & -1 & -1 \\ 0 & 1 & -1 \\ 0 & 0 & 1 \end{bmatrix}.$$

We have the solution $\mathbf{x} = B\mathbf{b}$. We write $B = A^{-1}$ and $\mathbf{x} = A^{-1}\mathbf{b}$. Note that we did not specify the values of $\mathbf{b}$. The system in question has a unique solution for every given $\mathbf{b} \in \mathbb{R}^3$. □

Example 1.3.4. Let $u = (1, 0, 0)$, $v = (1, 1, 0)$ and $w^* = (0, 1, 0)$. $\mathbf{b} = (b_1, b_2, b_3)$. We solve system
$$A\mathbf{x} = \mathbf{b},$$

where

$$A = [u \quad v \quad w^*] = \begin{bmatrix} 1 & 1 & 0 \\ 0 & 1 & 1 \\ 0 & 0 & 0 \end{bmatrix}, \ \mathbf{x} = \begin{bmatrix} x \\ y \\ z \end{bmatrix}.$$

That is, we solve

$$\begin{bmatrix} 1 & 1 & 0 \\ 0 & 1 & 1 \\ 0 & 0 & 0 \end{bmatrix} \begin{bmatrix} x \\ y \\ z \end{bmatrix} = \begin{bmatrix} b_1 \\ b_2 \\ b_3 \end{bmatrix}.$$

We notice that A is also a triangular matrix but we cannot solve the system by back substitution. The third equation of the system is

$$0 = b_3,$$

which may or may not be true depending on the value of b_3.

If $b_3 \neq 0$, system $A\mathbf{x} = \mathbf{b}$ has no solution.

If $b_3 = 0$, system $A\mathbf{x} = \mathbf{b}$ becomes

$$\begin{bmatrix} 1 & 1 & 0 \\ 0 & 1 & 1 \end{bmatrix} \begin{bmatrix} x \\ y \\ z \end{bmatrix} = \begin{bmatrix} b_1 \\ b_2 \end{bmatrix},$$

which has a free variable z that can be parameterized by $z = t$, $t \in \mathbb{R}$. Then we have

$$\begin{bmatrix} x \\ y \\ z \end{bmatrix} = \begin{bmatrix} b_1 - b_2 + t \\ b_2 - t \\ t \end{bmatrix} = \begin{bmatrix} b_1 - b_2 \\ b_2 \\ 0 \end{bmatrix} + t \begin{bmatrix} 1 \\ -1 \\ 1 \end{bmatrix}, \ t \in \mathbb{R}, \qquad (1.7)$$

which represents infinitely many solutions on a straight line in $\mathbb{R}^3$. □

Let us make some observations on the previous two examples. In Example 1.3.4, for arbitrary $\mathbf{b} \in \mathbb{R}^3$, we have a unique solution $x = A^{-1}\mathbf{b}$. That is, the vector equation

$$xu + yv + zw = \mathbf{b}$$

always has a unique solution for the linear combination coefficients (x, y, z). This implies that the set of vectors $\{u, v, w\}$ can span the whole space $\mathbb{R}^3$.

In Example 1.3.3, there exists $\mathbf{b} = (b_1, b_2, b_3) \in \mathbb{R}^3$ with $b_3 \neq 0$, which is not a linear combination of $\{u, v, w^*\}$. That is, the set of vectors $\{u, v, w^*\}$ cannot span the whole space $\mathbb{R}^3$. But why can $\{u, v, w\}$, while both sets have three different vectors? The answer is that $\{u, v, w^*\}$ has redundant vectors, but $\{u, v, w\}$ does not. Namely, the role of some vectors in $\{u, v, w^*\}$ can be replaced by other vectors. To identify the redundancy, we set up the following model:

$$x_1 u + x_2 v + x_3 w^* = 0,$$

solving for (x_1, x_2, x_3). By (1.7), we have at least one nonzero solution $(x_1, x_2, x_3) = (1, -1, 1)$. That is,

$$1u + (-1)v + 1\,w^* = 0 \Longleftrightarrow v = u + w^*.$$

That is, v can be replaced with $u + w^*$. Therefore, the spanning role of $\{u, v, w^*\}$ is the same as that of $\{u, w^*\}$, which cannot span a three dimensional space.

Next we verify that there is no redundancy in $\{u, v, w\}$ for spanning $\mathbb{R}^3$. We also set up the following model:

$$x_1 u + x_2 v + x_3 w = 0,$$

solving for (x_1, x_2, x_3). By the solution in Example 1.3.4, we have the only solution $(x_1, x_2, x_3) = (0, 0, 0)$. This means that none of the vectors in $\{u, v, w\}$ can be replaced by a linear combination of the other ones. They are **linearly independent**.

> **Definition 1.3.5.** Let $\{u_1, u_2, \cdots, u_m\}$ be a set of vectors in $\mathbb{R}^n$. If the vector equation
>
> $$x_1 u_1 + x_2 u_2 + \cdots + x_n u_n = 0$$
>
> has only the trivial solution $x_1 = x_2 = \cdots = x_n = 0$, $\{u_1, u_2, \cdots, u_m\}$ is said to be linearly independent. Otherwise, $\{u_1, u_2, \cdots, u_m\}$ is said to be linearly dependent.

We finish this chapter with examples on matrix multiplication with elementary matrices.

Example 1.3.6. Elementary matrices

- Consider $E\mathbf{x} = \mathbf{b}$, where $\mathbf{b} = \begin{bmatrix} b_1 \\ b_2 \end{bmatrix}$, $\mathbf{x} = \begin{bmatrix} x_1 \\ x_2 \end{bmatrix}$, $E = \begin{bmatrix} 1 & 0 \\ l & 1 \end{bmatrix}$. Then $E\mathbf{x} = \begin{bmatrix} x_1 \\ x_2 + lx_1 \end{bmatrix}$. Note that the effect of multiplication by E from the left of $\mathbf{x}$ is "adding l-multiple of row 1 to row 2." The solution is

$$\mathbf{x} = \begin{bmatrix} b_1 \\ b_2 - lb_1 \end{bmatrix}$$
$$= \begin{bmatrix} 1 & 0 \\ -l & 1 \end{bmatrix} \begin{bmatrix} b_1 \\ b_2 \end{bmatrix}.$$

Denote by $E^{-1} = \begin{bmatrix} 1 & 0 \\ -l & 1 \end{bmatrix}$. We have the solution $\mathbf{x} = E^{-1}\mathbf{b}$. The effect of multiplication by E^{-1} from the left of $\mathbf{b}$ is "subtracting l-multiple of row 1 from row 2." Moreover, using $\mathbf{x} = E^{-1}\mathbf{b}$ and the original system $E\mathbf{x} = \mathbf{b}$, we have

$$E(E^{-1}\mathbf{b}) = \mathbf{b}, \quad E^{-1}(E\mathbf{x}) = \mathbf{x}.$$

That is, the multiplication actions from the left of a vector by E and E^{-1} are canceling each other. If we treat the action $A : \mathbf{x} \mapsto A\mathbf{x}$ as a function determined by the matrix A, then the effect from $E^{-1} \circ E$ and $E \circ E^{-1}$ is the same as the identity matrix I.

- Consider $E\mathbf{x} = \mathbf{b}$, where $\mathbf{b} = \begin{bmatrix} b_1 \\ b_2 \end{bmatrix}$, $\mathbf{x} = \begin{bmatrix} x_1 \\ x_2 \end{bmatrix}$, $E = \begin{bmatrix} 0 & 1 \\ 1 & 0 \end{bmatrix}$. Then $E\mathbf{x} = \begin{bmatrix} x_2 \\ x_1 \end{bmatrix}$. Note that the effect of multiplication by E from the left of $\mathbf{x}$ is "exchanging positions of row 1 and row 2." The solution is

$$\mathbf{x} = \begin{bmatrix} b_2 \\ b_1 \end{bmatrix}$$
$$= \begin{bmatrix} 0 & 1 \\ 1 & 0 \end{bmatrix} \begin{bmatrix} b_1 \\ b_2 \end{bmatrix}.$$

Denote by $E^{-1} = \begin{bmatrix} 0 & 1 \\ 1 & 0 \end{bmatrix}$, which is identical to E itself. We have the solution $\mathbf{x} = E^{-1}\mathbf{b}$. The effect of multiplication by E^{-1} from the left of $\mathbf{b}$ is "exchanging positions of row 1 and row 2." Moreover, using $\mathbf{x} = E^{-1}\mathbf{b}$ and the original system $E\mathbf{x} = \mathbf{b}$, we have

$$E(E^{-1}\mathbf{b}) = \mathbf{b}, \quad E^{-1}(E\mathbf{x}) = \mathbf{x}.$$

That is, the multiplication actions from the left of a vector by E and E^{-1} are canceling each other. The multiplication effects from $E^{-1} \circ E$ and $E \circ E^{-1}$ are the same as the identity matrix I.

- Consider $E\mathbf{x} = \mathbf{b}$, where $\mathbf{b} = \begin{bmatrix} b_1 \\ b_2 \end{bmatrix}$, $\mathbf{x} = \begin{bmatrix} x_1 \\ x_2 \end{bmatrix}$, $E = \begin{bmatrix} 1 & 0 \\ 0 & c \end{bmatrix}$ with $c \neq 0$. Then $E x = \begin{bmatrix} x_1 \\ cx_2 \end{bmatrix}$. Note that the effect of multiplication by E from the left of $\mathbf{x}$ is "multiplying row 2 by c". The solution is

$$\mathbf{x} = \begin{bmatrix} b_1 \\ \frac{1}{c}b_2 \end{bmatrix}$$

$$= \begin{bmatrix} 1 & 0 \\ 0 & \frac{1}{c} \end{bmatrix} \begin{bmatrix} b_1 \\ b_2 \end{bmatrix}.$$

Denote by $E^{-1} = \begin{bmatrix} 1 & 0 \\ 0 & \frac{1}{c} \end{bmatrix}$. We have the solution $\mathbf{x} = E^{-1}\mathbf{b}$. The effect of multiplication by E^{-1} from the left of $\mathbf{b}$ is "dividing row 2 by c". Moreover, using $\mathbf{x} = E^{-1}\mathbf{b}$ and the original system $E\mathbf{x} = \mathbf{b}$, we have

$$E(E^{-1}\mathbf{b}) = \mathbf{b}, \quad E^{-1}(E\mathbf{x}) = \mathbf{x}.$$

That is, the multiplication actions from the left of a vector by E and E^{-1} are canceling each other. If we treat the action $A : \mathbf{x} \mapsto A\mathbf{x}$ as a function determined by the matrix A, then the effect from $E^{-1} \circ E$ and $E \circ E^{-1}$ is the same as the identity matrix I.

The aforementioned three type of matrices are called **elementary matrices** which can be obtained by operating on the identity matrices with the elementary row operation in question. □

Exercise 1.3.7.

1. Let $A = \begin{bmatrix} 1 & 0 & 0 \\ 1 & 1 & 0 \\ 1 & 1 & 1 \end{bmatrix}$ and $B = \begin{bmatrix} 1 & 0 & 0 \\ -1 & 1 & 0 \\ 0 & -1 & 1 \end{bmatrix}$. Compute i) $A + B$, $A + 2B$ and $A - 3B$; ii) AB and BA.

2. Let $A = \begin{bmatrix} 1 & 2 \\ 3 & 4 \end{bmatrix}$ and $B = \begin{bmatrix} 0 & 1 \\ 1 & 0 \end{bmatrix}$. i) Compute AB and BA; ii) Is $AB = BA$?

3. Find matrices A and b such that the system

$$\begin{cases} x + z = 1 \\ 2x + 5y + 8z = -1 \\ x + y = 1 \end{cases}$$

can be rewritten into matrix form $Ax = b$, where $x = (x, y, z)$.

4. Let $e_1 = \begin{bmatrix} 1 \\ 0 \\ 0 \end{bmatrix}$, $e_2 = \begin{bmatrix} 0 \\ 1 \\ 0 \end{bmatrix}$ and $e_3 = \begin{bmatrix} 0 \\ 0 \\ 1 \end{bmatrix}$. Determine that $\{e_1, e_2, e_3\}$ is a linearly independent set of vectors in $\mathbb{R}^3$.

5. Let $v_1 = \begin{bmatrix} 1 \\ 1 \\ 0 \end{bmatrix}$, $v_2 = \begin{bmatrix} 0 \\ 1 \\ 1 \end{bmatrix}$ and $v_3 = \begin{bmatrix} 1 \\ 0 \\ 1 \end{bmatrix}$. Determine whether or not $\{v_1, v_2, v_3\}$ is a linearly independent set of vectors in $\mathbb{R}^3$.

6. Let $u = (1, 0)$, $v = (1, 1)$ and $w = (1, 2)$. Show that $\{u, v, w\}$ is not linearly independent.

7. Let $S = \{u_1, u_2, \cdots, u_n\}$ be a set of vectors in $\mathbb{R}^n$. If one of them is the zero vector, is S linearly independent?

8. Let $\{v_1, v_2, v_3\} \subset \mathbb{R}^n$ be a set of linearly independent vectors. Determine whether $\{v_1 + v_2, v_2 + v_3, v_3 + v_1\}$ is linearly independent or not.

9. Show that every set of four vectors in $\mathbb{R}^3$ is linearly dependent.

10. Find the canceling matrices E^{-1} of the following E's.

$$E = \begin{bmatrix} 1 & 0 & 0 \\ 0 & 1 & 0 \\ 0 & 1 & 1 \end{bmatrix}, \quad E = \begin{bmatrix} 1 & 0 & 0 \\ 0 & 0 & 1 \\ 0 & 1 & 0 \end{bmatrix}, \quad E = \begin{bmatrix} 1 & 0 & 0 \\ 0 & 10 & 0 \\ 0 & 0 & 1 \end{bmatrix}.$$

Chapter 2

Solving linear systems

With the preparation of Chapter 1, we start the discussion of how to solve linear systems, using matrix representation of linear systems and matrix multiplications. In the process we certainly will develop related properties of matrices.

2.1   Vectors and linear equations

Solving a linear system $Ax = b$ means that we use certain operations on the system to reduce it into the form $\mathbf{x} = \mathbf{c}$, or equivalently $I\mathbf{x} = \mathbf{c}$, where I is the identity matrix. Such operations should be reversible, in the sense that the solution $\mathbf{x} = \mathbf{c}$ should be equivalent to the original system. By default, we agree on the fact that if $u = v$, u, $v \in \mathbb{R}^n$, then $Bu = Bv$ for every $m \times n$ matrix B. Using the language of matrix, to reduce $Ax = b$ into $I\mathbf{x} = \mathbf{c}$, we need to find a sequence of matrices E_1, E_2, $\cdots E_q$ such that

$$E_1 Ax = E_1 b \Rightarrow E_2 E_1 Ax = E_2 E_1 b \Rightarrow \cdots \Rightarrow E_q \cdots E_2 E_1 Ax = E_q \cdots E_2 E_1 b.$$

Namely, we keep multiplying both sides of the equation by the same matrix. The question is at which step we should stop. *If the solution is unique*, that is, the solution $\mathbf{x}$ must be a definite value, we should be able to arrive at the situation that the coefficient matrix of $\mathbf{x}$ becomes the identity matrix I. That is, $E_q \cdots E_2 E_1 A = I$. In order to make sure the solution $x = \mathbf{c}$ is equivalent to the original system, ideal candidates for the matrices E_1, E_2, $\cdots E_q$ are those elementary matrices, because we know their canceling matrices E_1^{-1}, E_2^{-1}, $\cdots E_q^{-1}$ such that we have

$$A\mathbf{x} = \mathbf{b} \Leftarrow E_1^{-1}E_1 A\mathbf{x} = E_1^{-1}E_1\mathbf{b} \Leftarrow \cdots \Leftarrow E_q^{-1}E_q \cdots E_1 A\mathbf{x} = E_q^{-1}E_q \cdots E_1\mathbf{b}.$$

Example 2.1.1.

We solve the following system $A\mathbf{x} = \mathbf{b}$, which is in system form:

$$\begin{cases} x - y = 1 \\ x + y = 2. \end{cases}$$

We follow our basic idea of eliminating variables, and at the same time use matrices to represent the elimination processes. We use the notation R_i to denote the i-th row or the i-th equation.

Solution:

System	Matrix representation	Elementary matrix
$\begin{cases} x - y = 1 \\ x + y = 2 \end{cases}$	$\begin{bmatrix} 1 & -1 \\ 1 & 1 \end{bmatrix}\begin{bmatrix} x \\ y \end{bmatrix} = \begin{bmatrix} 1 \\ 2 \end{bmatrix}$	
$\Downarrow R_2 + R_1$		$E_1 = \begin{bmatrix} 1 & 0 \\ 1 & 1 \end{bmatrix}$
$\begin{cases} x - y = 1 \\ \quad 2x = 3 \end{cases}$	$\begin{bmatrix} 1 & 0 \\ 1 & 1 \end{bmatrix}\begin{bmatrix} 1 & -1 \\ 1 & 1 \end{bmatrix}\begin{bmatrix} x \\ y \end{bmatrix} = \begin{bmatrix} 1 & 0 \\ 1 & 1 \end{bmatrix}\begin{bmatrix} 1 \\ 2 \end{bmatrix}$	
$\Downarrow R_2 \leftrightarrow R_1$	$\begin{bmatrix} 1 & -1 \\ 2 & 0 \end{bmatrix}\begin{bmatrix} x \\ y \end{bmatrix} = \begin{bmatrix} 1 \\ 3 \end{bmatrix}$	$E_2 = \begin{bmatrix} 0 & 1 \\ 1 & 0 \end{bmatrix}$
$\begin{cases} \quad 2x = 3 \\ x - y = 1 \end{cases}$	$\begin{bmatrix} 0 & 1 \\ 1 & 0 \end{bmatrix}\begin{bmatrix} 1 & -1 \\ 2 & 0 \end{bmatrix}\begin{bmatrix} x \\ y \end{bmatrix} = \begin{bmatrix} 0 & 1 \\ 1 & 0 \end{bmatrix}\begin{bmatrix} 1 \\ 3 \end{bmatrix}$	
$\Downarrow R_2 - \frac{1}{2}R_1$	$\begin{bmatrix} 2 & 0 \\ 1 & -1 \end{bmatrix}\begin{bmatrix} x \\ y \end{bmatrix} = \begin{bmatrix} 3 \\ 1 \end{bmatrix}$	$E_3 = \begin{bmatrix} 1 & 0 \\ -\frac{1}{2} & 1 \end{bmatrix}$
$\begin{cases} \quad 2x = 3 \\ -y = -\frac{1}{2} \end{cases}$	$\begin{bmatrix} 1 & 0 \\ -\frac{1}{2} & 1 \end{bmatrix}\begin{bmatrix} 2 & 0 \\ 1 & -1 \end{bmatrix}\begin{bmatrix} x \\ y \end{bmatrix} = \begin{bmatrix} 1 & 0 \\ -\frac{1}{2} & 1 \end{bmatrix}\begin{bmatrix} 3 \\ 1 \end{bmatrix}$	
$\Downarrow R_1 \cdot \frac{1}{2}$	$\begin{bmatrix} 2 & 0 \\ 0 & -1 \end{bmatrix}\begin{bmatrix} x \\ y \end{bmatrix} = \begin{bmatrix} 3 \\ -\frac{1}{2} \end{bmatrix}$	$E_4 = \begin{bmatrix} \frac{1}{2} & 0 \\ 0 & 1 \end{bmatrix}$
$\begin{cases} \quad x = \frac{3}{2} \\ -y = -\frac{1}{2} \end{cases}$	$\begin{bmatrix} \frac{1}{2} & 0 \\ 0 & 1 \end{bmatrix}\begin{bmatrix} 2 & 0 \\ 0 & -1 \end{bmatrix}\begin{bmatrix} x \\ y \end{bmatrix} = \begin{bmatrix} \frac{1}{2} & 0 \\ 0 & 1 \end{bmatrix}\begin{bmatrix} 3 \\ -\frac{1}{2} \end{bmatrix}$	
$\Downarrow R_2 \cdot (-1)$	$\begin{bmatrix} 1 & 0 \\ 0 & -1 \end{bmatrix}\begin{bmatrix} x \\ y \end{bmatrix} = \begin{bmatrix} \frac{3}{2} \\ -\frac{1}{2} \end{bmatrix}$	$E_5 = \begin{bmatrix} 1 & 0 \\ 0 & -1 \end{bmatrix}$

$$\begin{cases} x = \dfrac{3}{2} \\ y = \dfrac{1}{2} \end{cases} \quad \begin{bmatrix} 1 & 0 \\ 0 & -1 \end{bmatrix} \begin{bmatrix} 1 & 0 \\ 0 & -1 \end{bmatrix} \begin{bmatrix} x \\ y \end{bmatrix} = \begin{bmatrix} 1 & 0 \\ 0 & -1 \end{bmatrix} \begin{bmatrix} \frac{3}{2} \\ -\frac{1}{2} \end{bmatrix}$$

$$\begin{bmatrix} 1 & 0 \\ 0 & 1 \end{bmatrix} \begin{bmatrix} x \\ y \end{bmatrix} = \begin{bmatrix} \frac{3}{2} \\ \frac{1}{2} \end{bmatrix}.$$

Indeed, in the process we have $E_5 E_4 E_3 E_2 E_1 A = I$ and the solution is obtained when an identity matrix appears in the last step. Let us also observe that the so far undefined matrix multiplication between 2×2 matrices actually can be done **column by column on the second matrix**. (Please verify it in the computation.) For general matrices, we will follow this convention which will formally become our definition of matrix multiplication — we are actually justifying how convenient it is for solving linear systems.

Let us observe that during the elimination process in Example 2.1.1, if the first step was $R_2 - R_1$, then interchanging of two rows in the second step could have been avoided. The upper leftmost nonzero entry 1 in row 1 is called a **pivot** or a leading 1. One could use this pivot to eliminate all entries below it. If we have a pivot in each row during the elimination, we can reduce every nonpivot entry of a square matrix into zeros. □

Notice that the variables (x, y) in the matrix representation are actually unnecessarily carried in each step. We use the so-called **augmented matrix**, which is the coupling of the coefficient matrix and the right hand side of the system, to represent the system by a single matrix, completely dropping (x, y).

Example 2.1.2. Solve system (1.1):

$$\begin{cases} x + 2y + 3z = 3 \\ 2x + 5y + 8z = 9 \\ 3x + 6y + 18z = 18. \end{cases}$$

Solution: We re-write the system of linear equations in the matrix form $A\mathbf{x} = \mathbf{b}$, where

$$A = \begin{bmatrix} 1 & 2 & 3 \\ 2 & 5 & 8 \\ 3 & 6 & 18 \end{bmatrix}, \ \mathbf{x} = \begin{bmatrix} x \\ y \\ z \end{bmatrix} \text{ and } \mathbf{b} = \begin{bmatrix} 3 \\ 9 \\ 18 \end{bmatrix}.$$

Then the corresponding augmented matrix is

$$[A : \mathbf{b}] = \begin{bmatrix} 1 & 2 & 3 & 3 \\ 2 & 5 & 8 & 9 \\ 3 & 6 & 18 & 18 \end{bmatrix}.$$

By the elementary row operations on $[A : \mathbf{b}]$ we have

$$\begin{bmatrix} 1 & 2 & 3 & 3 \\ 2 & 5 & 8 & 9 \\ 3 & 6 & 18 & 18 \end{bmatrix} \xrightarrow{R_2 - 2R_1} \begin{bmatrix} 1 & 2 & 3 & 3 \\ 0 & 1 & 2 & 3 \\ 3 & 6 & 18 & 18 \end{bmatrix}$$

$$\xrightarrow{R_3+(-3)R_1} \begin{bmatrix} 1 & 2 & 3 & 3 \\ 0 & 1 & 2 & 3 \\ 0 & 0 & 9 & 9 \end{bmatrix}$$

$$\xrightarrow{R_3/9} \begin{bmatrix} 1 & 2 & 3 & 3 \\ 0 & 1 & 2 & 3 \\ 0 & 0 & 1 & 1 \end{bmatrix}$$

$$\xrightarrow{R_2-2R_3} \begin{bmatrix} 1 & 2 & 3 & 3 \\ 0 & 1 & 0 & 1 \\ 0 & 0 & 1 & 1 \end{bmatrix}$$

$$\xrightarrow{R_1-3R_3} \begin{bmatrix} 1 & 2 & 0 & 0 \\ 0 & 1 & 0 & 1 \\ 0 & 0 & 1 & 1 \end{bmatrix}$$

$$\xrightarrow{R_1-2R_2} \begin{bmatrix} 1 & 0 & 0 & -2 \\ 0 & 1 & 0 & 1 \\ 0 & 0 & 1 & 1 \end{bmatrix}.$$

Then we have an equivalent system with augmented matrix

$$\begin{bmatrix} 1 & 0 & 0 & -2 \\ 0 & 1 & 0 & 1 \\ 0 & 0 & 1 & 1 \end{bmatrix}.$$

The solution is

$$\begin{bmatrix} x \\ y \\ z \end{bmatrix} = \begin{bmatrix} -2 \\ 1 \\ 1 \end{bmatrix}.$$

□

We remark that in the solution of Example 2.1.2, the coefficient matrix has been reduced into an upper triangular matrix (the entries below the main diagonal are all zeros) after two eliminations. Once an upper triangular matrix is obtained, we can use back substitution to solve for z, and y and x. The remaining steps are eliminating the entries above the pivots so that we obtain an equivalent system with a diagonal/identity coefficient matrix whose solution will be directly displayed.

The following example deals with the situation that *the solution is not unique*, but we still carry out the elimination process until we arrive at the situation that the maximal number of variables has coefficients 1.

Example 2.1.3. Solve the following system of linear equations.

$$\begin{cases} x_1 + x_2 + x_3 + 2x_4 = -1 \\ 2x_1 + x_2 + 3x_3 + x_4 = -2 \\ 2x_1 - 2x_2 + 4x_3 + 2x_4 = -3. \end{cases}$$

Solution: We re-write the system of linear equations in the matrix form $A\mathbf{u} = \mathbf{b}$, where

$$A = \begin{bmatrix} 1 & 1 & 1 & 2 \\ 2 & 1 & 3 & 1 \\ 2 & -2 & 4 & 2 \end{bmatrix}, \ \mathbf{u} = \begin{bmatrix} x_1 \\ x_2 \\ x_3 \\ x_4 \end{bmatrix} \text{ and } \mathbf{b} = \begin{bmatrix} -1 \\ -2 \\ -3 \end{bmatrix}.$$

Then the corresponding augmented matrix is

$$[A : b] = \begin{bmatrix} 1 & 1 & 1 & 2 & -1 \\ 2 & 1 & 3 & 1 & -2 \\ 2 & -2 & 4 & 2 & -3 \end{bmatrix}.$$

By the elementary row operations on $[A : b]$ we have

$$\begin{bmatrix} 1 & 1 & 1 & 2 & -1 \\ 2 & 1 & 3 & 1 & -2 \\ 2 & -2 & 4 & 2 & -3 \end{bmatrix} \xrightarrow[R_2 - 2R_1]{R_3 + (-2)R_1} \begin{bmatrix} 1 & 1 & 1 & 2 & -1 \\ 0 & -1 & 1 & -3 & 0 \\ 0 & 4 & 2 & -2 & -1 \end{bmatrix}$$

$$\xrightarrow{R_2(-1)} \begin{bmatrix} 1 & 1 & 1 & 2 & -1 \\ 0 & 1 & -1 & 3 & 0 \\ 0 & -4 & 2 & -2 & -1 \end{bmatrix}$$

$$\xrightarrow[R_3 + (4)R_2]{R_1 - R_2} \begin{bmatrix} 1 & 0 & 2 & -1 & -1 \\ 0 & 1 & -1 & 3 & 0 \\ 0 & 0 & -2 & 10 & -1 \end{bmatrix}$$

$$\xrightarrow{R_3/(-2)} \begin{bmatrix} 1 & 0 & 2 & -1 & -1 \\ 0 & 1 & -1 & 3 & 0 \\ 0 & 0 & 1 & -5 & \frac{1}{2} \end{bmatrix}$$

$$\xrightarrow[R_2 + R_3]{R_1 - 2R_3} \begin{bmatrix} 1 & 0 & 0 & 9 & -2 \\ 0 & 1 & 0 & -2 & \frac{1}{2} \\ 0 & 0 & 1 & -5 & \frac{1}{2} \end{bmatrix}.$$

Then we have a system with augmented matrix

$$\begin{bmatrix} 1 & 0 & 0 & 9 & -2 \\ 0 & 1 & 0 & -2 & \frac{1}{2} \\ 0 & 0 & 1 & -5 & \frac{1}{2} \end{bmatrix}.$$

Let $x_4 = t$, where t is an arbitrary real number. The solution is

$$\begin{bmatrix} x_1 \\ x_2 \\ x_3 \\ x_4 \end{bmatrix} = \begin{bmatrix} -2 - 9t \\ \frac{1}{2} + 2t \\ \frac{1}{2} + 5t \\ t \end{bmatrix}, \ t \in \mathbb{R}.$$

□

Notice that the reduced system does not have an identity coefficient matrix, even if it is close to it. We say such a matrix is in **reduced row echelon form**, which is a matrix satisfying the following:

> **i)** If a row is not entirely zero, the first nonzero entry is 1, which we call the pivot 1 or leading 1;
>
> **ii)** The zero rows are exchanged to the bottom;
>
> **iii)** For every two pivot 1's (leading 1's), the one in the higher row is closer from the left of the matrix;
>
> **iv)** Each column has a leading 1 and has zeros everywhere else in the column.

A matrix satisfying conditions i), ii) and iii) is said to be in row echelon form. The elimination process to obtain a row echelon form is called **Gaussian elimination**. The elimination process to obtain a reduced row echelon form is called **Gauss–Jordan elimination**.

Example 2.1.4. Consider matrices

$$A = \begin{bmatrix} 1 & 0 & 0 & 9 & -2 \\ 0 & 0 & 1 & -5 & \frac{1}{2} \\ 0 & 0 & 0 & 1 & 4 \end{bmatrix}, B = \begin{bmatrix} 1 & 0 & 0 & 9 & -2 \\ 0 & 0 & 0 & 1 & 4 \\ 0 & 0 & 1 & -5 & \frac{1}{2} \end{bmatrix}.$$

A is in row echlon form, but B is not because the pivot in row 2 is farther from the left of the matrix than the pivots in row 3. The pivots should be positioned in the matrix in a staircase shape.

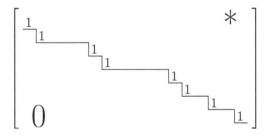

□

The following example illustrates how we determine a system has a unique solution, has no solution or has infinitely many solutions.

Example 2.1.5. Let k be a real number. Consider the following linear system of equations:

$$\begin{cases} x_2 + x_3 + x_4 = -1 \\ 2x_1 + x_2 + 3x_3 + x_4 = -2 \\ 2x_1 - 2x_2 + 4x_3 + 2x_4 = -3 \\ 3x_2 - x_3 - x_4 = k. \end{cases} \tag{2.1}$$

Find all possible values of k such that system (2.1) has a unique solution, has no solutions and has infinitely many solutions.

Solution: We re-write the system of linear equations in the matrix form $A\mathbf{u} = \mathbf{b}$, where

$$A = \begin{bmatrix} 0 & 1 & 1 & 1 \\ 2 & 1 & 3 & 1 \\ 2 & -2 & 4 & 2 \\ 0 & 3 & -1 & -1 \end{bmatrix}, \ \mathbf{u} = \begin{bmatrix} x_1 \\ x_2 \\ x_3 \\ x_4 \end{bmatrix} \text{ and } \mathbf{b} = \begin{bmatrix} -1 \\ -2 \\ -3 \\ k \end{bmatrix}.$$

Then the corresponding augmented matrix is

$$[A : \mathbf{b}] = \begin{bmatrix} 0 & 1 & 1 & 1 & -1 \\ 2 & 1 & 3 & 1 & -2 \\ 2 & -2 & 4 & 2 & -3 \\ 0 & 3 & -1 & -1 & k \end{bmatrix}.$$

By the elementary row operations on $[A : \mathbf{b}]$ we have

$$\begin{bmatrix} 0 & 1 & 1 & 1 & -1 \\ 2 & 1 & 3 & 1 & -2 \\ 2 & -2 & 4 & 2 & -3 \\ 0 & 3 & -1 & -1 & k \end{bmatrix}$$

$$\xrightarrow{R_3 \leftrightarrow R_2} \begin{bmatrix} 2 & 1 & 3 & 1 & -2 \\ 0 & 1 & 1 & 1 & -1 \\ 2 & -2 & 4 & 2 & -3 \\ 0 & 3 & -1 & -1 & k \end{bmatrix}$$

$$\xrightarrow[R_4 - 3R_2]{R_3 + (-1)R_1} \begin{bmatrix} 2 & 1 & 3 & 1 & -2 \\ 0 & 1 & 1 & 1 & -1 \\ 0 & -3 & 1 & 1 & -1 \\ 0 & 0 & -4 & -4 & k+3 \end{bmatrix}$$

$$\xrightarrow[R_3 + 3R_2]{R_1 - R_2} \begin{bmatrix} 2 & 0 & 2 & 0 & -1 \\ 0 & 1 & 1 & 1 & -1 \\ 0 & 0 & 4 & 4 & -4 \\ 0 & 0 & -4 & -4 & k+3 \end{bmatrix}$$

$$\xrightarrow{R_4 + R_3} \begin{bmatrix} 2 & 0 & 2 & 0 & -1 \\ 0 & 1 & 1 & 1 & -1 \\ 0 & 0 & 4 & 4 & -4 \\ 0 & 0 & 0 & 0 & k-1 \end{bmatrix}$$

$$\xrightarrow{R_3/4}
\begin{bmatrix}
2 & 0 & 2 & 0 & -1 \\
0 & 1 & 1 & 1 & -1 \\
0 & 0 & 1 & 1 & -1 \\
0 & 0 & 0 & 0 & k-1
\end{bmatrix}$$

$$\xrightarrow[R_1/2]{R_2-R_3}
\begin{bmatrix}
1 & 0 & 1 & 0 & -\frac{1}{2} \\
0 & 1 & 0 & 0 & 0 \\
0 & 0 & 1 & 1 & -1 \\
0 & 0 & 0 & 0 & k-1
\end{bmatrix}$$

$$\xrightarrow{R_1-R_3}
\begin{bmatrix}
1 & 0 & 0 & -1 & \frac{1}{2} \\
0 & 1 & 0 & 0 & 0 \\
0 & 0 & 1 & 1 & -1 \\
0 & 0 & 0 & 0 & k-1
\end{bmatrix}.$$

Then we arrive at

$$\begin{bmatrix}
1 & 0 & 0 & -1 & \frac{1}{2} \\
0 & 1 & 0 & 0 & 0 \\
0 & 0 & 1 & 1 & -1 \\
0 & 0 & 0 & 0 & k-1
\end{bmatrix},$$

which can immediately lead to the reduced echelon form of the augmented matrix $[A : \mathbf{b}]$ if we know the value of k.

i) If $k \neq 1$, then the system is not consistent and has no solution because the last equation is $0 = 1$ which is contradictory.

ii) If $k = 1$, then the system is consistent and has infinitely many solutions. Let $x_4 = t$, where t is an arbitrary real number. The solution is

$$\begin{bmatrix} x_1 \\ x_2 \\ x_3 \\ x_4 \end{bmatrix} = \begin{bmatrix} \frac{1}{2} + t \\ 0 \\ -1 - t \\ t \end{bmatrix}, t \in \mathbb{R}.$$

□

Exercise 2.1.6.

1. Redo Example 2.1.1 with the first elementary row operation $R_2 - R_1$.

2. Determine whether the following matrices are in reduced row echelon form and row echelon form, respectively:

$$\begin{bmatrix}
1 & 0 & 0 & 9 & -2 \\
0 & 1 & 0 & -2 & \frac{1}{2} \\
0 & 0 & 1 & -5 & \frac{1}{2}
\end{bmatrix}, \quad
\begin{bmatrix}
1 & 0 & 0 & 9 \\
0 & 1 & 0 & 1 \\
0 & 0 & 1 & -5
\end{bmatrix}, \quad
\begin{bmatrix}
1 & 1 & 0 & 1 \\
0 & 1 & 0 & \frac{1}{2} \\
0 & 0 & 1 & \frac{1}{2}
\end{bmatrix}.$$

3. Solve the following systems using Gauss–Jordan eliminations:

a) $\begin{cases} x + 3z = 1 \\ 2x + 3y = 3 \\ 4y + 5z = 5 \end{cases}$ b) $\begin{cases} x + 2y + 3z = 1 \\ 2x + 3y + 4z = 3 \\ 3x + 4y + 5z = 5 \end{cases}$ c) $\begin{cases} x + 2y + 3z = 1 \\ 2x + 3y + 4z = 3 \\ 5x + 9y + 13z = 7 \end{cases}$

4. Consider a linear system $Ax = b$ with A an $m \times n$ matrix and b an $m \times 1$ matrix. Is it true if there is more than one solution for x in $\mathbb{R}^n$, there must be infinitely many? You may use the fact that

$$A(x + y) = Ax + Ay$$
$$A(tx) = tAx,$$

for every $x, y \in \mathbb{R}^n$ and $t \in \mathbb{R}$,

which is called the **linearity** of matrix multiplication.

5. Let k be a real number. Consider the following linear system of equations:

$$\begin{cases} x_2 + 2x_3 + x_4 = 1 \\ 2x_1 + x_2 + 3x_3 = 2 \\ x_1 + 4x_3 + 2x_4 = 3 \\ kx_2 + x_4 = 1. \end{cases} \tag{2.2}$$

Find all possible values of k such that system (2.2) i) has a unique solution; ii) has no solutions and iii) has infinitely many solutions.

2.2   Matrix operations

We have dealt with matrix multiplication when we represent systems of linear equations into matrix form $Ax = b$, where Ax is a $m \times n$ matrix times an $n \times 1$ matrix. In this section, we discuss matrix operations for general matrices.

Definition 2.2.1. Let A, B be $m \times n$ matrices, $c \in \mathbb{R}$ a scalar. Then $A + B$ and cA are $m \times n$ matrices defined by

$$(A + B)_{ij} = (A)_{ij} + (B)_{ij}, \ (cA)_{ij} = c(A)_{ij},$$

where $(A)_{ij}$ denotes the entry at the (ij) position of A.

Example 2.2.2. Let $A = \begin{bmatrix} 1 & 2 & 3 \\ 3 & 4 & 5 \end{bmatrix}$, $B = \begin{bmatrix} 5 & 4 & 3 \\ 3 & 2 & 1 \end{bmatrix}$. Then we have

$$A + B = \begin{bmatrix} 6 & 6 & 6 \\ 6 & 6 & 6 \end{bmatrix}, \quad 2A = \begin{bmatrix} 2 & 4 & 6 \\ 6 & 8 & 10 \end{bmatrix}.$$

$\square$

Recall that in Example 2.1.1, we multiplied square matrices with the convention that the first matrix multiplies column by column on the second matrix of the product AB. One point to note is that a product $A\mathbf{x}$ with a column matrix $\mathbf{x}$ exists if the number of columns of A equals the number of rows of $\mathbf{x}$.

We have

Definition 2.2.3. Let A be an $m \times n$ matrix, B be an $n \times r$ matrix. Then AB is an $m \times r$ matrix.

$$AB = [Ab_1 : Ab_2 : \cdots : Ab_r],$$

where $B = [b_1 : b_2 : \cdots : b_r]$.

Example 2.2.4. $A = \begin{bmatrix} 1 & 2 \\ 3 & 4 \end{bmatrix}$, $B = \begin{bmatrix} 5 & 4 & 3 \\ 3 & 2 & 1 \end{bmatrix}$. Then we have

$$AB = \begin{bmatrix} 11 & 8 & 5 \\ 27 & 20 & 13 \end{bmatrix}.$$

$\square$

Let A be an $m \times n$ matrix, B be an $n \times r$ matrix. If we partition A into rows and B into columns, we have

$$AB = \begin{bmatrix} a_1 \\ a_2 \\ \vdots \\ a_m \end{bmatrix} \begin{bmatrix} b_1 : b_2 : \cdots : b_r \end{bmatrix}$$

$$= \begin{bmatrix} a_1b_1 & a_1b_2 & \cdots & a_1b_n \\ a_2b_1 & a_2b_2 & \cdots & a_2b_n \\ \cdots & \cdots & \cdots & \cdots \\ a_mb_1 & a_mb_2 & \cdots & a_mb_r \end{bmatrix},$$

from which we have

$$AB = \begin{bmatrix} a_1B \\ a_2B \\ \vdots \\ a_mB \end{bmatrix},$$

and that

$$(AB)_{i,j} = (\text{Row } i \text{ of } A) \cdot (\text{Column } j \text{ of } B).$$

We have the following associative property of matrix products.

Theorem 2.2.5. Let A, B and C be $m \times n$, $n \times p$ and $p \times q$ matrices, respectively. Then we have

$$(AB)C = A(BC).$$

Proof. First of all we know that both products are $m \times q$. For every position (i, j) in the product matrix, we have

$$((AB)C)_{ij} = (\text{Row } i \text{ of } AB) \cdot (\text{Column } j \text{ of } C)$$

$$= \sum_{s=1}^{p} (AB)_{is} C_{sj}$$

$$= \sum_{s=1}^{p} [(\text{Row } i \text{ of } A) \cdot (\text{Column } s \text{ of } B)] C_{sj}$$

$$= \sum_{s=1}^{p} \left[\sum_{t=1}^{n} A_{it} B_{ts} \right] C_{sj}$$

$$= \sum_{t=1}^{n} A_{it} \left[\sum_{s=1}^{p} B_{ts} C_{sj} \right]$$

$$= (\text{Row } i \text{ of } A) \cdot (\text{Column } j \text{ of } BC)$$

$$= (A(BC))_{ij}.$$

$\square$

By the approach of comparing corresponding positions in the related matrices, we have the following distributive properties:

Lemma 2.2.6. Let A and B be $m \times n$; C and D be $n \times p$ matrices; $t \in \mathbb{R}$ be a scalar. We have

$A(C + D) = AC + AD$;

$(A + B)C = AC + BC$;

$t(A + B) = tA + tB$.

Unfortunately there is no commutative property in general for matrix products.

In general,
$$AB \neq BA.$$

Example 2.2.7. Let $A = \begin{bmatrix} 1 & 2 \\ 3 & 4 \end{bmatrix}$, $B = \begin{bmatrix} 0 & 1 \\ 1 & 0 \end{bmatrix}$. We have

$$AB = \begin{bmatrix} 2 & 1 \\ 4 & 3 \end{bmatrix}, \quad BA = \begin{bmatrix} 3 & 4 \\ 1 & 2 \end{bmatrix},$$

which illustrates that $AB \neq BA$. Observe that B is an elementary matrix which exchanges rows of A if multiplied on the left of A, but exchanges columns of A if multiplied on the right. We will understand this after we have learned transposition in the later sections. $\square$

We finish this section with the definition of matrix powers:

$$A^m = \underbrace{AAA \cdots A}_{m \text{ copies of } A}.$$

Exercise 2.2.8.

1. Let $A = \begin{bmatrix} 1 & 2 \\ 3 & 4 \end{bmatrix}$, $B = \begin{bmatrix} 1 & 0 & 1 & 0 \\ 0 & 1 & 0 & 1 \end{bmatrix}$. i) Compute AB. ii) Does BA exist?

2. Let $A = \begin{bmatrix} 1 & 2 \\ 3 & 4 \end{bmatrix}$, $B = \begin{bmatrix} 1 & -1 & 0 & 1 \\ 1 & 1 & 1 & 0 \end{bmatrix}$. i) Compute AB. ii) If B is block partitioned into $B = [B_1 : B_2]$, is it true $AB = [AB_1 : AB_2]$?

3. Show Lemma 2.2.6.

4. Let $A = [a_1 : a_2 \cdots : a_n]$, $B = \begin{bmatrix} b_1 \\ b_2 \\ \vdots \\ b_n \end{bmatrix}$ be $m \times n$ and $n \times r$ matrices. Show that $AB = a_1 b_1 + a_2 b_2 + \cdots + a_n b_n$.

5. Let $A = \begin{bmatrix} 1 & 2 \\ 3 & 4 \end{bmatrix}$, $B = \begin{bmatrix} 1 & -1 & 0 & 1 \\ 1 & 1 & 1 & 0 \end{bmatrix}$. Use Question 4 to compute AB.

6. Let A and B be $m \times n$ and $n \times r$ matrices. Show that i) every column of AB is a linear combination of the columns of A; ii) every row of AB is a linear combination of the rows of B.

7. Let $A = \begin{bmatrix} 1 & 2 \\ 3 & 4 \end{bmatrix}$. Find all matrices B such that $AB = BA$.

8. Let A and B be $n \times n$ matrices. Explain that in general we have $(A - B)(A - B) \neq A^2 - B^2$ and $(A + B)^2 \neq A^2 + 2AB + B^2$.

9. Let A be an $n \times n$ matrix. Define $V = \{B : AB = BA\}$. Show that i) $V \neq \emptyset$; ii) if $B_1 \in V$ and $B_2 \in V$, then every linear combination of B_1 and B_2 is in V.

10. Give an example that $A^2 = 0$ but $A \neq 0$.

11. Give an example that $A^2 = I$ but $A \neq \pm I$.

12. Let A be an $n \times n$ matrix. If we want to define a limit $\lim_{m \to \infty} A^m$, how would you define the closeness (distance) between matrices?

2.3 Inverse matrices

To solve the linear system $Ax = b$ where A is an $n \times n$ matrix and b is an arbitrary $n \times 1$ vector, we use the elimination process to reduce it into $Ix = x_0$, where I is the $n \times n$ identity matrix. We know each step of elimination can be represented by a left multiplication of an elementary matrix on both sides of the current linear system. That is,

$$E_m E_{m-1} \cdots E_2 E_1 Ax = E_m E_{m-1} \cdots E_2 E_1 b.$$

If the eliminations reduce A into the identity matrix I, that is, $E_m E_{m-1} \cdots E_2 E_1 A = I$, we obtain the solution $x = E_m E_{m-1} \cdots E_2 E_1 Ab$. For notational convenience, we write $B = E_m E_{m-1} \cdots E_2 E_1$. We have

$$BA = I \text{ and } x = Bb.$$

If we bring $x = Bb$ back into the original system, we have

$$ABb = b.$$

Since b is assumed to be arbitrary, we have $ABb = b$ for every b. Then we have
$$AB = I.$$

In summary, in order to have a unique solution for $Ax = b$ with arbitrary b, we need

$$BA = I = AB. \tag{2.3}$$

We notice that

> **Lemma 2.3.1.** If there exists an $n \times n$ matrix B satisfying $BA = I = AB$, then it is unique.

This is because if $CA = I = AC$ we have

$$C = CI = C(AB) = (CA)B = IB = B.$$

> **Definition 2.3.2.** Let A be an $n \times n$ matrix. If there exists B such that $BA = I = AB$, A is said to be invertible. B is called the **inverse** of A and is denoted by A^{-1}. If there is no such matrix B satisfying $AB = I = BA$, A is said to be singular, or not invertible.

An immediate application of the definition of inverse is that if a matrix has a zero row or zero column, it cannot be invertible. Say A has a row of zeros, then by matrix multiplication, for every matrix B such that AB exists, AB will have a row of zeros, which cannot be equal to the identity matrix.

> If A has a zero row or zero column, A is not invertible.

Example 2.3.3.   1. Since $I^2 = II = I$, the identity matrix is invertible with $I^{-1} = I$.

2. Let $A = \text{diag}\{d_1, d_2, \cdots, d_n\}$. If $d_1 d_2 \cdots d_n \neq 0$, then A is invertible with $A^{-1} = \text{diag}\{d_1^{-1}, d_2^{-1}, \cdots, d_n^{-1}\}$. To be more specific, if $d_1 d_2 \cdots d_n \neq 0$, then we have

$$
A = \begin{bmatrix} d_1 & & & \\ & d_2 & & \\ & & \ddots & \\ & & & d_n \end{bmatrix} \implies A^{-1} = \begin{bmatrix} d_1^{-1} & & & \\ & d_2^{-1} & & \\ & & \ddots & \\ & & & d_n^{-1} \end{bmatrix}.
$$

3. All elementary matrices are invertible. For instance,

$$
E_{31} = \begin{bmatrix} 1 & 0 & 0 \\ 0 & 1 & 0 \\ 2 & 0 & 1 \end{bmatrix} \implies E_{31}^{-1} = \begin{bmatrix} 1 & 0 & 0 \\ 0 & 1 & 0 \\ -2 & 0 & 1 \end{bmatrix},
$$

$$
P_{21} = \begin{bmatrix} 0 & 1 & 0 \\ 1 & 0 & 0 \\ 0 & 0 & 1 \end{bmatrix} \implies P_{21}^{-1} = \begin{bmatrix} 0 & 1 & 0 \\ 1 & 0 & 0 \\ 0 & 0 & 1 \end{bmatrix} = P_{21},
$$

$$
E_{33} = \begin{bmatrix} 1 & 0 & 0 \\ 0 & 1 & 0 \\ 0 & 0 & 10 \end{bmatrix} \implies E_{33}^{-1} = \begin{bmatrix} 1 & 0 & 0 \\ 0 & 1 & 0 \\ 0 & 0 & \frac{1}{10} \end{bmatrix}.
$$

□

Now we show two equivalent conditions for invertibility of a matrix.

> **Theorem 2.3.4.** The following are equivalent:
>
>   i) A is invertible.
>
>   ii) $Ax = 0$ has only the trivial solution $x = 0$.
>
>   iii) A is equal to a product of elementary matrices.

Proof. i) $\Rightarrow$ ii) Since A is invertible, the inverse A^{-1} exists. Then $Ax = 0$ leads to $A^{-1}Ax = A^{-1}0$ and $Ix = x = 0$.

ii) $\Rightarrow$ iii) If $Ax = 0$ has only the trivial solution $x = 0$, then by Gaussian elimination, the augmented matrix $[A : 0]$ can be reduced to $[I : 0]$. That is, there exist elementary matrices $E_1, E_2, \cdots, E_m$ such that

$$E_m E_{m-1} \cdots E_2 E_1 [A : 0] = [I : 0].$$

Hence $E_m E_{m-1} \cdots E_2 E_1 A = I$ and we have $A = E_1^{-1} E_2^{-1} \cdots E_m^{-1}$. Since $E_i^{-1}, i = 1, 2, \cdots, m$ are again elementary matrices, A is a product of elementary matrices.

iii) $\Rightarrow$ i). Suppose A is a product of elementary matrices with $A = E_1^{-1} E_2^{-1} \cdots E_m^{-1}$. Let $B = E_m E_{m-1} \cdots E_2 E_1$. We have

$$
\begin{aligned}
BA &= (E_m E_{m-1} \cdots E_2 E_1)(E_1^{-1} E_2^{-1} \cdots E_m^{-1}) \\
&= I \\
&= (E_1^{-1} E_2^{-1} \cdots E_m^{-1})(E_m E_{m-1} \cdots E_2 E_1) \\
&= AB.
\end{aligned}
$$

$\square$

Theorem 2.3.5. Let A and B be $n \times n$ matrices.

i) If $BA = I$, then $B = A^{-1}$.

ii) If $AB = I$, then $B = A^{-1}$.

Proof. i) If A is invertible, we immediately have $B = B(AA^{-1}) = (BA)A^{-1} = A^{-1}$. To show A is invertible by Theorem 2.3.4, we show that $Ax = 0$ has only the trivial solution. Indeed, if $Ax = 0$, we have $BAx = B0$ which implies that $Ix = x = 0$. $Ax = 0$ has only the trivial solution. Hence A is invertible.

ii) By 1), we have $A = B^{-1}$ and B is invertible. Notice that $BB^{-1} = B^{-1}B = I$. B^{-1} is invertible with inverse equal to B. Therefore, A is invertible with $A^{-1} = (B^{-1})^{-1} = B$. That is, $B = A^{-1}$. $\square$

Theorem 2.3.6. Let A and B be $n \times n$ matrices. Then A and B are invertible if and only if AB is invertible.

Proof. "$\Longrightarrow$" If A and B are invertible, A^{-1} and B^{-1} exist. Then we have

$$(B^{-1}A^{-1})(AB) = (AB)(B^{-1}A^{-1}) = I,$$

which imply that AB is invertible.

"$\Longleftarrow$" If AB is invertible, then there exists a matrix C such that $C(AB) = (AB)C = I$. By associative property of matrix product, we have

$$(CA)B = A(BC) = I,$$

in which by Theorem 2.3.5 A and B are both invertible. $\square$

> **Lemma 2.3.7.** Let $A = \begin{bmatrix} a & b \\ c & d \end{bmatrix}$.
>
> A is invertible if and only if $ad - bc \neq 0$.
>
> If A is invertible, then
>
> $$A^{-1} = \frac{1}{ad - bc} \begin{bmatrix} d & -b \\ -c & a \end{bmatrix}.$$

Proof. We show the first part. The second part can be directly verified by computing AA^{-1}.

"$\Longrightarrow$" If A is invertible, it cannot have a zero row. Therefore, c and d cannot be simultaneously zero. Suppose $ad - bc = 0$, then the system $A\mathbf{x} = \mathbf{0}$ has a nontrivial solution

$$\mathbf{x} = \begin{bmatrix} d \\ -c \end{bmatrix} \neq \mathbf{0}.$$

By Theorem 2.3.4, A is not invertible. This is a contradiction. Hence $ad - bc \neq 0$.

"$\Longleftarrow$" If $ad - bc \neq 0$, the matrix

$$B = \frac{1}{ad - bc} \begin{bmatrix} d & -b \\ -c & a \end{bmatrix}$$

exists and satisfies $BA = I$. Therefore, A is invertible with $A^{-1} = B = \frac{1}{ad-bc} \begin{bmatrix} d & -b \\ -c & a \end{bmatrix}$. □

From Lemma 2.3.7, we know that the invertibility is determined by a scalar quantity $ad - bc$. We call it the **determinant** of A. In the later chapters we will come back to this notion.

Example 2.3.8. (Dominant matrices are invertible) We call an $n \times n$ matrix $A = (a_{ij})$ a **dominant matrix** if for every $i \in \{1, 2, \cdots, n\}$, we have

$$|a_{ii}| > \sum_{j \neq i} |a_{ij}|.$$

We show that if A is dominant, then it is invertible.

Proof. We show that if $\mathbf{x} \neq 0$ then $A\mathbf{x} \neq 0$ so that Theorem 2.3.4 applies. Let $|x_{i_0}|$ be the largest coordinate of $\mathbf{x}$ in absolute value. Then we have

$$\begin{aligned}
|(Ax)_{i_0 1}| &= |(\text{Row } i_0 \text{ of } A) \cdot \mathbf{x}| \\
&= |a_{i_0 1}x_1 + a_{i_0 2}x_2 + \cdots + a_{i_0 i_0}x_{i_0} + \cdots + a_{i_0 n}x_n| \\
&\geq |a_{i_0 i_0}x_{i_0}| - |a_{i_0 1}x_1 + a_{i_0 2}x_2 + \cdots + a_{i_0(i_0-1)}x_{i_0-1} \\
&\quad + a_{i_0(i_0+1)}x_{i_0+1} \cdots + a_{i_0 n}x_n|
\end{aligned}$$

$$\geq |a_{i_0 i_0} x_{i_0}| - |a_{i_0 1} x_1| - |a_{i_0 2} x_2| - \cdots - |a_{i_0 (i_0 -1)} x_{i_0 -1}|$$
$$- |a_{i_0 (i_0 +1)} x_{i_0 +1}| \cdots - |a_{i_0 n} x_n|$$
$$\geq |a_{i_0 i_0} x_{i_0}| - |a_{i_0 1} x_{i_0}| - |a_{i_0 2} x_{i_0}| - \cdots - |a_{i_0 (i_0 -1)} x_{i_0}|$$
$$- |a_{i_0 (i_0 +1)} x_{i_0}| \cdots - |a_{i_0 n} x_0|$$
$$= \left(|a_{i_0 i_0}| - \sum_{j \neq i_0} |a_{i_0 j}| \right) |x_{i_0}| > 0.$$

$\square$

We finish this section by a concrete example on how to find the inverse of a square matrix, if it exists. The idea is the same as solving the linear system by elimination. If we can find elementary matrices $E_1, E_2, \cdots, E_m$ such that $E_m E_{m-1} \cdots E_1 A = I$, then we obtain $A^{-1} = E_m E_{m-1} \cdots E_1$. The problem is that it is not economical that we compute the matrix product after we have found all of the elementary matrices. We must find a device to record the product at the same time of elimination. Indeed the coupled matrix $[A : I]$ serves this purpose very well. Namely, we have

$$A^{-1}[A : I] = [A^{-1}A : A^{-1}] = [I : A^{-1}].$$

The inverse A^{-1} is recorded at the second part of the coupled matrix when the first part becomes I.

Example 2.3.9. For the given matrix A, we use elimination to find A^{-1} and record elementary row operation and the corresponding elementary matrix at the same time.

$$A = \begin{bmatrix} 3 & 0 & 7 \\ 2 & 1 & 4 \\ 1 & -1 & 2 \end{bmatrix}.$$

Solution:

$$\begin{bmatrix} 3 & 0 & 7 & 1 & 0 & 0 \\ 2 & 1 & 4 & 0 & 1 & 0 \\ 1 & -1 & 2 & 0 & 0 & 1 \end{bmatrix}$$

	Row operation	Elementary Matrix
$\Downarrow$	$R_1 \leftrightarrow R_3$	$E_1 = \begin{bmatrix} 0 & 0 & 1 \\ 0 & 1 & 0 \\ 1 & 0 & 0 \end{bmatrix}$

$$\begin{bmatrix} 1 & -1 & 2 & 0 & 0 & 1 \\ 2 & 1 & 4 & 0 & 1 & 0 \\ 3 & 0 & 7 & 1 & 0 & 0 \end{bmatrix}$$

	Row operation	Elementary Matrix
$\Downarrow$	$R_2 - 2R_1$	$E_2 = \begin{bmatrix} 1 & 0 & 0 \\ -2 & 1 & 0 \\ 0 & 0 & 1 \end{bmatrix}$

$$\begin{bmatrix} 1 & -1 & 2 & 0 & 0 & 1 \\ 0 & 3 & 0 & 0 & 1 & -2 \\ 3 & 0 & 7 & 1 & 0 & 0 \end{bmatrix}$$

Row operation	Elementary Matrix
$R_3 - 3R_1$	$E_3 = \begin{bmatrix} 1 & 0 & 0 \\ 0 & 1 & 0 \\ -3 & 0 & 1 \end{bmatrix}$

$$\Downarrow$$

$$\begin{bmatrix} 1 & -1 & 2 & 0 & 0 & 1 \\ 0 & 3 & 0 & 0 & 1 & -2 \\ 0 & 3 & 1 & 1 & 0 & -3 \end{bmatrix}$$

Row operation	Elementary Matrix
$R_3 - R_2$	$E_4 = \begin{bmatrix} 1 & 0 & 0 \\ 0 & 1 & 0 \\ 0 & -1 & 1 \end{bmatrix}$

$$\Downarrow$$

$$\begin{bmatrix} 1 & -1 & 2 & 0 & 0 & 1 \\ 0 & 3 & 0 & 0 & 1 & -2 \\ 0 & 0 & 1 & 1 & -1 & -1 \end{bmatrix}$$

Row operation	Elementary Matrix
$R_2/3$	$E_5 = \begin{bmatrix} 1 & 0 & 0 \\ 0 & \frac{1}{3} & 0 \\ 0 & 0 & 1 \end{bmatrix}$

$$\Downarrow$$

$$\begin{bmatrix} 1 & -1 & 2 & 0 & 0 & 1 \\ 0 & 1 & 0 & 0 & \frac{1}{3} & -\frac{2}{3} \\ 0 & 0 & 1 & 1 & -1 & -1 \end{bmatrix}$$

Row operation	Elementary Matrix
$R_1 + R_2$	$E_6 = \begin{bmatrix} 1 & 1 & 0 \\ 0 & 1 & 0 \\ 0 & 0 & 1 \end{bmatrix}$

$$\Downarrow$$

$$\begin{bmatrix} 1 & 0 & 2 & 0 & \frac{1}{3} & \frac{1}{3} \\ 0 & 1 & 0 & 0 & \frac{1}{3} & -\frac{2}{3} \\ 0 & 0 & 1 & 1 & -1 & -1 \end{bmatrix}$$

Row operation	Elementary Matrix
$R_1 - 2R_3$	$E_7 = \begin{bmatrix} 1 & 0 & -2 \\ 0 & 1 & 0 \\ 0 & 0 & 1 \end{bmatrix}$

$$\Downarrow$$

$$\begin{bmatrix} 1 & 0 & 0 & -2 & \frac{7}{3} & \frac{7}{3} \\ 0 & 1 & 0 & 0 & \frac{1}{3} & -\frac{2}{3} \\ 0 & 0 & 1 & 1 & -1 & -1 \end{bmatrix}$$

Then we have

$$A^{-1} = \begin{bmatrix} -2 & \frac{7}{3} & \frac{7}{3} \\ 0 & \frac{1}{3} & -\frac{2}{3} \\ 1 & -1 & -1 \end{bmatrix},$$

which can be written as the product of elementary matrices $E_7 E_6 E_5 E_4 E_3 E_2 E_1$.

□

Exercise 2.3.10.

1. Determine whether or not the following matrices are invertible. Find the inverse of each matrix if it exists.

$$a) \begin{bmatrix} 1 & 2 \\ 3 & 4 \end{bmatrix}, \quad b) \begin{bmatrix} -1 & 2 \\ 3 & 6 \end{bmatrix}, \quad c) \begin{bmatrix} 1 & 2 \\ 3 & 6 \end{bmatrix}.$$

2. Determine whether or not the following matrices are invertible. Find the inverse of each matrix if it exists.

$$a) \begin{bmatrix} 0 & 0 & 1 \\ 0 & 1 & 0 \\ 1 & 0 & 0 \end{bmatrix}, \quad b) \begin{bmatrix} 1 & 0 & 0 \\ 2 & 1 & 0 \\ 0 & 3 & 1 \end{bmatrix}, \quad c) \begin{bmatrix} 1 & 0 & 0 \\ 0 & 6 & 0 \\ 0 & 0 & 1 \end{bmatrix}.$$

3. Determine whether or not the following matrices are invertible. Find the inverse of each matrix if it exists.

$$u) \begin{bmatrix} 1 & 2 & 0 \\ 3 & 4 & 0 \\ 0 & 0 & 1 \end{bmatrix}, \quad b) \begin{bmatrix} 1 & 0 & 0 \\ 0 & -1 & 2 \\ 0 & 3 & 6 \end{bmatrix}, \quad c) \begin{bmatrix} 1 & 2 & 0 \\ 3 & 6 & 0 \\ 0 & 0 & 1 \end{bmatrix}.$$

4. For the given matrix A, use elimination to find A^{-1} and record each elementary row operation and the corresponding elementary matrix at the same time.

$$A = \begin{bmatrix} 3 & 0 & 1 \\ 2 & 4 & 2 \\ 1 & -1 & 5 \end{bmatrix}.$$

5. For what values of $\lambda \in \mathbb{R}$ is the following matrix

$$A = \begin{bmatrix} 3 & 0 & 1 \\ 2 & 4 & 2 \\ 1 & -1 & \lambda \end{bmatrix}$$

invertible?

6. Let A be an $n \times n$ matrix. If $A = \begin{bmatrix} r_1 \\ r_2 \\ \vdots \\ r_n \end{bmatrix}$ satisfies that $r_2 = r_3 + r_1$, is A invertible?

7. Let A be an $n \times n$ matrix. If $A = \begin{bmatrix} c_1 : c_2 : \cdots : c_n \end{bmatrix}$ satisfies that $c_2 = c_3 + c_1$, is A invertible?

8. Let $v, w \in \mathbb{R}^n$ be vectors. Is the matrix $A = \begin{bmatrix} \|v\| & 1 \\ |v \cdot w| & \|w\| \end{bmatrix}$ invertible?

9. Give an example of a 3×3 dominant matrix and find its inverse.

10. Find a sufficient condition on a, b, c and $d \in \mathbb{R}$ such that the matrix

$$A = \begin{bmatrix} a^2 + b^2 & 2ab \\ 2cd & c^2 + d^2 \end{bmatrix}$$

is invertible.

11. Let $A = \begin{bmatrix} 0 & 1 & 1 \\ 0 & 0 & 1 \\ 0 & 0 & 0 \end{bmatrix}$. i) Compute A^2; ii) Show that for every $k \geq 3$, $k \in \mathbb{N}$, $A^k = 0$.

12. Let A be an $n \times n$ matrix. Show that if $A^k = 0$, then $I - A$ is invertible and

$$(I - A)^{-1} = I + A + A^2 + \cdots + A^{k-1}.$$

13. Let A be an $n \times n$ matrix and $A = tI + N$, $t \in \mathbb{R}$ with $N^4 = 0$ for some $k \in \mathbb{N}$. Compute A^4 in terms of t and N.

14. Let $D = \text{diag}\{\lambda_1, \lambda_2, \cdots, \lambda_n\}$ be a diagonal matrix with the main diagonal entries λ_1, λ_2, $\cdots$, λ_n. Show that D is invertible if and only if $\lambda_i \neq 0$, for every $i = 1, 2, \cdots, n$.

15. Let A be an $n \times n$ matrix. i) If $A^3 = I$, find A^{-1}; ii) If $A^k = I$ for some $k \in \mathbb{N}$, find A^{-1}; iii) If $A^k = 0$ for some $k \in \mathbb{N}$, is it possible that A is invertible?

16. Show that A is invertible if and only if A^k is invertible for every $k \in \mathbb{N}, k \geq 1$.

17. Let A and B be $n \times n$ invertible matrices. i) Give an example to show that $A + B$ may not be invertible; ii) Show that $A + B$ is invertible if and only if $A^{-1} + B^{-1}$ is invertible.

2.4   *LU* decomposition

We have observed in solving $A\mathbf{x} = \mathbf{b}$ with A an $n \times n$ matrix that once we have a triangular coefficient matrix, say $U\mathbf{x} = \mathbf{c}$, we can use back substitution to solve the system. Indeed, if elementary matrices $E_1, E_2, \cdots, E_m$ are lower triangular and reduce A into an upper triangular matrix without row exchanges, then the product $L = E_1^{-1}E_2^{-1}\cdots E_m^{-1}$ is lower triangular. In such a case, $A\mathbf{x} = \mathbf{b}$ is equivalent to

$$LU\mathbf{x} = \mathbf{b},$$

and we can solve the system by solving two systems with triangular matrices:

$$L\mathbf{c} = \mathbf{b}, \quad U\mathbf{x} = \mathbf{c},$$

which can be solved by forward and backward substitution, respectively.

Next we explain that L can be obtained without really carrying out the matrix multiplications. Assume that each elementary matrix in E_1, E_2, $\cdots$, E_m deals with a different position (i, j) of an $n \times n$ matrix with $i > j$ and we assume that the elimination on A was carried out with the order E_1, $E_2, \cdots, E_m$, successively on A. Then the product $L = E_1^{-1}E_2^{-1} \cdots E_m^{-1} = E_1^{-1}E_2^{-1} \cdots E_m^{-1}I$ is acting on I backward adding the $-l_{i,j}$ multiple of the j-th row to the i-th row, where $l_{i,j}$ is the multiplier from $E_{i,j}$. With $p < q$, the row with the position to be altered by E_p^{-1} is not altered (relative to the identity matrix) when E_q^{-1} acts. To be specific, let us examine the following example:

Example 2.4.1. Let

$$A = \begin{bmatrix} 2 & -1 & 0 \\ -1 & 2 & -1 \\ 0 & -1 & 2 \end{bmatrix}, \; E_{21} = \begin{bmatrix} 1 & 0 & 0 \\ \frac{1}{2} & 1 & 0 \\ 0 & 0 & 1 \end{bmatrix}, \; E_{32} = \begin{bmatrix} 1 & 0 & 0 \\ 0 & 1 & 0 \\ 0 & \frac{2}{3} & 1 \end{bmatrix}.$$

We have

$$E_{32}E_{21}A = U = \begin{bmatrix} 2 & -1 & 0 \\ 0 & \frac{3}{2} & -1 \\ 0 & 0 & \frac{4}{3} \end{bmatrix}.$$

Then the LU decomposition is $A = LU$ with

$$L = E_{21}^{-1}E_{32}^{-1} = \begin{bmatrix} 1 & 0 & 0 \\ -\frac{1}{2} & 1 & 0 \\ 0 & 0 & 1 \end{bmatrix} \begin{bmatrix} 1 & 0 & 0 \\ 0 & 1 & 0 \\ 0 & -\frac{2}{3} & 1 \end{bmatrix} = \begin{bmatrix} 1 & 0 & 0 \\ -\frac{1}{2} & 1 & 0 \\ 0 & -\frac{2}{3} & 1 \end{bmatrix}.$$

We see that the multipliers $-l_{3,2} = -\frac{2}{3}$ and $-l_{2,1} = -\frac{1}{2}$ of the elementary matrices are placed directly into the identity matrix to form the product L. Row 2 for the position $(2, 1)$ which will be altered by E_{21}^{-1} is not altered when E_{32}^{-1} sends $-l_{3,2}$ to the $(3, 2)$-position. When E_{21}^{-1} acts after E_{32}^{-1}, it sends $-l_{2,1}$ to the $(2, 1)$-position using unchanged Row 1 to produce a new Row 2, but this new Row 2 has no effect on E_{32}^{-1} anymore.

If we compute

$$E_{32}^{-1}E_{21}^{-1} = \begin{bmatrix} 1 & 0 & 0 \\ -\frac{1}{2} & 1 & 0 \\ \frac{1}{3} & -\frac{2}{3} & 1 \end{bmatrix},$$

E_{21}^{-1} produces a nontrivial entry at $(2, 1)$ in Row 2. When E_{32}^{-1} acts, it uses an already altered Row 2 and produces an unwanted entry at $(3, 1)$. In summary, we have

If E_1, E_2, $\cdots$, E_m are lower triangular elementary matrices which operate on distinct positions (i, j), $i > j$ of the $n \times n$ matrix A and are such that $E_m E_{m-1} \cdots E_2 E_1 A = U$ is an upper triangular matrix, then $A = LU$, where $L = E_1^{-1} E_2^{-1} \cdots E_m^{-1}$ is lower triangular with

$$(L)_{ij} = -l_{ij},$$

and l_{ij} is the multiplier of the corresponding elementary matrix in E_1, E_2, $\cdots$, E_m which operates on position (i, j).

$\square$

Exercise 2.4.2.

1. Let $-l_{ij}$ be the entry of the 4×4 E_{ij}^{-1} matrix below the main diagonal. Which one of the following products can be obtained by directly writing $-l_{ij}$ into the (i, j) position of the products? i) $E_{31}^{-1} E_{32}^{-1} E_{41}^{-1} E_{42}^{-1} E_{43}^{-1}$; ii) $E_{32}^{-1} E_{21}^{-1} E_{31}^{-1} E_{42}^{-1} E_{43}^{-1}$.

2. Find the LU decomposition of

$$A = \begin{bmatrix} 3 & 0 & 1 \\ 2 & 4 & 2 \\ 1 & -1 & 5 \end{bmatrix}.$$

3. Let $b = (1, 2, 3)$ and $A = \begin{bmatrix} 3 & 0 & 1 \\ 2 & 4 & 2 \\ 1 & -1 & 5 \end{bmatrix}$. Use the LU decomposition of A to solve system $Ax = b$.

4. Is it true that a matrix A does not have an LU decomposition? Justify your answer.

2.5 Transpose and permutation

Definition 2.5.1. Let A be an $m \times n$ matrix. The transpose of A is an $n \times m$ matrix denoted A^T and is defined by

$$(A^T)_{ij} = (A)_{ji}.$$

Example 2.5.2. We have the following examples.

1. If $A = \begin{bmatrix} 1 & 2 & 3 \end{bmatrix}$, then $A^T = \begin{bmatrix} 1 \\ 2 \\ 3 \end{bmatrix}$.

2. If $A = \begin{bmatrix} 1 & 2 & 3 \\ 4 & 5 & 6 \\ 7 & 8 & 9 \end{bmatrix}$, then $A^T = \begin{bmatrix} 1 & 4 & 7 \\ 2 & 5 & 8 \\ 3 & 6 & 9 \end{bmatrix}$.

3. If $x, y \in \mathbb{R}^n$ are treated as $n \times 1$ matrices, then $x \cdot y = x^T y$.

$\square$

Theorem 2.5.3. Suppose that A and B are matrices such that $A+B$, AB and A^{-1} exist. We have

$$(A + B)^T = A^T + B^T, \tag{2.4}$$
$$(AB)^T = B^T A^T, \tag{2.5}$$
$$(A^{-1})^T = (A^T)^{-1}. \tag{2.6}$$

We show (2.5) while the others can be proved similarly. For every (i, j) position of $(AB)^T$ we have

$$\begin{aligned} ((AB)^T))_{ij} &= (AB)_{ji} \\ &= (\text{Row } j \text{ of } A) \cdot (\text{Column } i \text{ of } B) \\ &= (\text{Column } i \text{ of } B) \cdot (\text{Row } j \text{ of } A) \\ &= (\text{Row } i \text{ of } B^T) \cdot (\text{Column } j \text{ of } A^T) \\ &= (B^T A^T)_{ij}. \end{aligned}$$

Next we show (2.6). We need only to show $A^T (A^{-1})^T = I$, which is by (2.5) immediately true since $A^T (A^{-1})^T = (A^{-1}A)^T = I^T = I$.

Definition 2.5.4. Let A be an $n \times n$ matrix. If $A = A^T$, A is called a **symmetric matrix**.

Example 2.5.5. 1. Diagonal matrices are symmetric.

2. If $A = \begin{bmatrix} 1 & 2 & 3 \\ 2 & 5 & 4 \\ 3 & 4 & 6 \end{bmatrix}$, then we have $A^T = A$ and A is symmetric.

3. If A is an $m \times n$ matrix, then both AA^T and $A^T A$ are square matrices and are symmetric.

4. If $x, y \in \mathbb{R}^n$ are treated as $n \times 1$ matrices, then xy^T is an $n \times n$ matrix, but NOT symmetric in general.

$\square$

Theorem 2.5.6. Let A be an $m \times n$ matrix. Then $A^T A$ is invertible if and only if the columns of A are linearly independent.

Proof. "$\Longrightarrow$" If the columns of A are not linearly independent, then the system $Ax = 0$, which is equivalent to

$$Ax = [c_1 : c_2 : \cdots : c_n] \begin{bmatrix} x_1 \\ x_2 \\ \vdots \\ x_n \end{bmatrix} = 0 \Leftrightarrow x_1 c_1 + x_2 c_2 + \cdots x_n c_n = 0,$$

has nontrivial solutions. Then $A^T A x = 0$ also has nontrivial solutions since $Ax = 0$ implies $A^T A x = A^T 0 = 0$. By Theorem 2.3.4, $A^T A$ is not invertible. This is a contradiction.

"$\Longleftarrow$" Consider $A^T A x = 0$. We note that

$Ax \perp$ every column of A (or every row of A^T);

Ax itself is a linear combination of columns of A (or every row of A^T).

Ax must be orthogonal to itself. Hence $Ax = 0$. Since the columns of A are linearly independent, $x = 0$ is the only solution of $A^T A x = 0$. By Theorem 2.3.4, $A^T A$ is invertible. $\qquad\square$

LU decomposition of symmetric matrices

Recall that if an $n \times n$ matrix A has n pivots (nonzero), its LU decomposition can be written as $A = LDU$ where D is the diagonal matrix whose main diagonal entries are the pivots. Then $L^{-1} A = DU$, where L^{-1} is actually the product of all elementary matrices (without row exchange matrix) which reduce A into the upper triangular matrix DU.

Assume that A is symmetric. Then the lower triangular part (below the main diagonal) of A is the same as that of $(DU)^T$ — noting that DU is the remaining upper triangular part of A after elimination! Therefore, by the same set of eliminations, we can reduce $(DU)^T$ into D. That is

$$L^{-1}(DU)^T = D,$$

which is equivalent to $(L^{-1} U^T - I)D = 0$. Since the diagonal matrix D has all pivots at the main diagonal which are all nonzero, we have

$$L^{-1} U^T - I = 0.$$

That is, $L = U^T$. To summarize, we have

> If A is an $n \times n$ symmetric matrix with n pivots and no row exchange needed to have the decomposition $A = LDU$, then we have $U = L^T$ and
> $$A = LDL^T.$$

Example 2.5.7. Let $A = \begin{bmatrix} 1 & 2 & 3 \\ 2 & 5 & 6 \\ 3 & 6 & 7 \end{bmatrix}$. Then A is a symmetric matrix. We have

$$\begin{bmatrix} 1 & 2 & 3 \\ 2 & 5 & 6 \\ 3 & 6 & 7 \end{bmatrix} \xrightarrow{R_2 - 2R_1} \begin{bmatrix} 1 & 2 & 3 \\ 0 & 1 & 0 \\ 3 & 6 & 7 \end{bmatrix}$$

$$\xrightarrow{R_3 - 3R_1} \begin{bmatrix} 1 & 2 & 3 \\ 0 & 1 & 0 \\ 0 & 0 & -2 \end{bmatrix} = \begin{bmatrix} 1 & 0 & 0 \\ 0 & 1 & 0 \\ 0 & 0 & -2 \end{bmatrix} \begin{bmatrix} 1 & 2 & 3 \\ 0 & 1 & 0 \\ 0 & 0 & 1 \end{bmatrix} = DU,$$

where D and U denote the diagonal and upper triangular matrices in the last equality, respectively. We have $E_{31} E_{21} A = DU$, where

$$E_{21} = \begin{bmatrix} 1 & 0 & 0 \\ -2 & 1 & 0 \\ 0 & 0 & 1 \end{bmatrix}, \; E_{31} = \begin{bmatrix} 1 & 0 & 0 \\ 0 & 1 & 0 \\ -3 & 0 & 1 \end{bmatrix},$$

with

$$E_{21}^{-1} = \begin{bmatrix} 1 & 0 & 0 \\ 2 & 1 & 0 \\ 0 & 0 & 1 \end{bmatrix}, \; E_{31}^{-1} = \begin{bmatrix} 1 & 0 & 0 \\ 0 & 1 & 0 \\ 3 & 0 & 1 \end{bmatrix}.$$

We have $L = E_{21}^{-1} E_{31}^{-1} = \begin{bmatrix} 1 & 0 & 0 \\ 2 & 1 & 0 \\ 3 & 0 & 1 \end{bmatrix}$. Then the LDL^T decomposition of A is

$$A = \begin{bmatrix} 1 & 0 & 0 \\ 2 & 1 & 0 \\ 3 & 0 & 1 \end{bmatrix} \begin{bmatrix} 1 & 0 & 0 \\ 0 & 1 & 0 \\ 0 & 0 & -2 \end{bmatrix} \begin{bmatrix} 1 & 2 & 3 \\ 0 & 1 & 0 \\ 0 & 0 & 1 \end{bmatrix}.$$

□

Remark 2.5.8. So far we have avoided an exchange of rows for LU decomposition. However, there are indeed cases where an exchange of rows is necessary to reduce a matrix into upper triangular form. For instance, the following elimination has to have a row exchange to obtain an upper triangular form:

$$A = \begin{bmatrix} 1 & 2 & 3 \\ 4 & 8 & 10 \\ 0 & 1 & 0 \end{bmatrix} \implies \begin{bmatrix} 1 & 2 & 3 \\ 0 & 0 & -2 \\ 0 & 1 & 0 \end{bmatrix}.$$

In this example if we do know in advance row 2 and row 3 should be exchanged to have a LU decomposition, we could decompose $P_{32}A$ such that $P_{32}A = LDU$.

> If A is invertible, then there exists a permutation matrix P such that $PA = LDU$ where L and U are lower and upper triangular matrices, respectively.

Permutations

Definition 2.5.9. An $n \times n$ matrix P is called a **permutation matrix** if the identity matrix can be obtained by rearranging the rows of P.

By definition, there are $n!$ permutation matrices of order n.

Example 2.5.10. Consider 3×3 permutation matrices.

$$I = \begin{bmatrix} 1 & 0 & 0 \\ 0 & 1 & 0 \\ 0 & 0 & 1 \end{bmatrix}, \ P_{21} = \begin{bmatrix} 0 & 1 & 0 \\ 1 & 0 & 0 \\ 0 & 0 & 1 \end{bmatrix}, \ P_{31} = \begin{bmatrix} 0 & 0 & 1 \\ 0 & 1 & 0 \\ 1 & 0 & 0 \end{bmatrix}, \ P_{32} = \begin{bmatrix} 1 & 0 & 0 \\ 0 & 0 & 1 \\ 0 & 1 & 0 \end{bmatrix},$$

$$P_{31}P_{32} = \begin{bmatrix} 0 & 1 & 0 \\ 0 & 0 & 1 \\ 1 & 0 & 0 \end{bmatrix}, \ P_{21}P_{32} = \begin{bmatrix} 0 & 0 & 1 \\ 1 & 0 & 0 \\ 0 & 1 & 0 \end{bmatrix}, \ P_{21}P_{31} = \begin{bmatrix} 0 & 1 & 0 \\ 0 & 0 & 1 \\ 1 & 0 & 0 \end{bmatrix}.$$

Notice that a permutation matrix P which represents a single permutation is symmetric and the inverse is itself. That is, $P^{-1} = P = P^T$. Let us call it a simple permutation matrix. For a general permutation matrix E which represents multiple permutations, it is a product of simple permutation matrices. Assume that $E = P_1 P_2 P_3 \cdots P_n$, where P_i, $i = 1, 2, \cdots, n$ are simple permutation matrices. Then we have

$$\begin{aligned} E^T &= (P_1 P_2 P_3 \cdots P_n)^T \\ &= P_n^T P_{n-1}^T \cdots P_1^T \\ &= P_n P_{n-1} \cdots P_1, \end{aligned} \tag{2.7}$$

which is not equal to E anymore. Therefore E may NOT be symmetric. However, by (2.7) we have

$$\begin{aligned} E^{-1} &= (P_1 P_2 P_3 \cdots P_n)^{-1} \\ &= P_n^{-1} P_{n-1}^{-1} \cdots P_1^{-1} \\ &= P_n P_{n-1} \cdots P_1 \\ &= E^T. \end{aligned}$$

That is, if E is a permutation matrix, then $EE^T = I = E^T E$ which imply that

$$(\text{Row } i \text{ of } E) \cdot (\text{Row } j \text{ of } E) = \begin{cases} 1 & \text{if } i = j, \\ 0 & \text{if } i \neq j, \end{cases}$$

and

$$(\text{Column } i \text{ of } E) \cdot (\text{Column } j \text{ of } E) = \begin{cases} 1 & \text{if } i = j, \\ 0 & \text{if } i \neq j. \end{cases}$$

The rows (columns) of E are orthogonal to each other. □

Definition 2.5.11. If A satisfies that $A^{-1} = A^T$, we call it an **orthogonal matrix**.

> If P is a permutation matrix, it is an orthogonal matrix. That is, $P^{-1} = P^T$.

Example 2.5.12. Let $A = \begin{bmatrix} 1 & 2 & 3 \\ 2 & 5 & 6 \\ 3 & 6 & 7 \end{bmatrix}$, and $P = \begin{bmatrix} 0 & 1 & 0 \\ 0 & 0 & 1 \\ 1 & 0 & 0 \end{bmatrix}$. Then A is symmetric and P is a nonsymmetric permutation matrix. So PA is a permutation of the rows of A, row 1 to row 3, row 3 to row 2 and row 2 to row 1. That is

$$PA = \begin{bmatrix} 2 & 5 & 6 \\ 3 & 6 & 7 \\ 1 & 2 & 3 \end{bmatrix}.$$

Since 1, 5, 7 have to be on the main diagonal, in order to restore symmetry from PA, column 1 has to be placed in column 3, column 3 goes to column 2 and column 2 to column 1. We know that a permutation matrix that multiplies from the right of a matrix will manipulate the columns. The aforementioned operations can be achieved by a right multiplication of

$$Q = \begin{bmatrix} 0 & 0 & 1 \\ 1 & 0 & 0 \\ 0 & 1 & 0 \end{bmatrix}.$$

That is, $PAQ = \begin{bmatrix} 5 & 6 & 2 \\ 6 & 7 & 3 \\ 2 & 3 & 1 \end{bmatrix}$. Note that $Q^T = P$ and PAP^T is always symmetric if A is symmetric. Therefore, if the symmetry of a matrix is destroyed by multiplication of a permutation matrix, we can restore symmetry from the product with a multiplication of its transpose from the other side. □

Example 2.5.13. Let $P = \begin{bmatrix} 0 & 0 & 1 & 0 \\ 1 & 0 & 0 & 0 \\ 0 & 0 & 0 & 1 \\ 0 & 1 & 0 & 0 \end{bmatrix}$. Then P is a permutation matrix and hence the inverse can be obtained by taking the transpose of P:

$$P^{-1} = P^T = \begin{bmatrix} 0 & 1 & 0 & 0 \\ 0 & 0 & 0 & 1 \\ 1 & 0 & 0 & 0 \\ 0 & 0 & 1 & 0 \end{bmatrix}.$$

Exercise 2.5.14.

1. Let $A = \begin{bmatrix} 0 & 1 & 0 & 0 \\ 0 & 0 & 1 & 0 \\ 0 & 0 & 0 & 1 \\ 1 & 0 & 0 & 0 \end{bmatrix}$. Find A^{-1} and A^T.

2. Let $A = \begin{bmatrix} 0 & 1 & 0 \\ 0 & 0 & 1 \\ 0 & 0 & 0 \\ 1 & 0 & 0 \end{bmatrix}$. i) Find AA^T and $A^T A$. ii) Determine which one of AA^T and $A^T A$ is invertible. iii) If one of AA^T and $A^T A$ is invertible, does it contradict Theorem 2.3.6?

3. Let
$$A = \begin{bmatrix} 0 & 1 & 2 & 3 \\ 1 & 2 & 3 & 4 \\ 2 & 3 & 4 & 5 \\ 3 & 4 & 5 & 6 \end{bmatrix}, \quad B = \begin{bmatrix} 1 & 2 & 3 & 4 \\ 0 & 1 & 2 & 3 \\ 3 & 4 & 5 & 6 \\ 2 & 3 & 4 & 5 \end{bmatrix}.$$
i) Find a permutation matrix P_1 such that $B = P_1 A$; ii) Find a permutation matrix P_2 such that $A = P_2 B$. iii) Compute $P_1 P_2$ and $P_2 P_1$.

4. Let
$$A = \begin{bmatrix} 7 & 1 & 2 & 3 \\ 1 & 2 & 3 & 4 \\ 2 & 3 & 4 & 5 \\ 3 & 4 & 5 & 7 \end{bmatrix}.$$
Find a permutation matrix P, a lower triangular matrix L and a diagonal matrix D such that $A = LDL^T$.

5. Let $R_\theta = \begin{bmatrix} \cos\theta & -\sin\theta \\ \sin\theta & \cos\theta \end{bmatrix}$. Show that R_θ is an orthogonal matrix.

6. Let $x \in \mathbb{R}^n$ with $x^T x = 1$. Define the **Householder matrix** by
$$H = I - 2xx^T.$$
i) Show that H is an orthogonal matrix; ii) Show that H is symmetric.

7. Let $S = \begin{bmatrix} I & A \\ A^T & O \end{bmatrix}$, where I is $m \times m$ and A is $m \times n$, O the zero matrix. Find a block diagonal matrix D and block lower triangular matrix L such that
$$S = LDL^T.$$

8. Show that AA^T is invertible if and only if the rows of A are linearly independent.

9. We say A is **skew-symmetric** if $A^T = -A$. i) Show that if A is a skew-symmetric $n \times n$ matrix then $a_{ii} = 0$ for every $i = 1, 2, \cdots, n$. ii) If A is both symmetric and skew-symmetric, then $A = 0$.

10. Let A be an $n \times n$ matrix. Show that i) $A + A^T$ is symmetric; ii) $A - A^T$ is skew-symmetric; iii) For every square matrix B, there exist a unique symmetric matrix B_1 and a unique skew-symmetric matrix B_2 such that $B = B_1 + B_2$.

11. A matrix is called **lower triangular** if every entry above the main diagonal is zero and is called **upper triangular** if every entry below the main diagonal is zero. Let A be an $n \times n$ invertible matrix. i) Show that if A is lower triangular, A^{-1} is also lower triangular; ii) Show that if A is upper triangular, then A^{-1} is also upper triangular.

Chapter 3

Vector spaces

3.1   Spaces of vectors

We know that the operations addition and scalar multiplication in Euclidean space $\mathbb{R}^n$ produce vectors within $\mathbb{R}^n$. Namely $\mathbb{R}^n$ is closed under addition and scalar multiplication. In addition, the derived operations, such as exchange of order of addition, do not produce a vector different from the one before the operation. To be specific, for u, v, $w \in \mathbb{R}^n$, s, $t \in \mathbb{R}$, $\mathbb{R}^n$ satisfies the following properties:

i) (Commutative property) $u + v = v + u$;

ii) (Associative property) $(u + v) + w = u + (v + u)$;

iii) (Identity element of addition) $u + 0 = u = 0 + u$;

iv) (Existence of addition inverse) $u + (-u) = 0$;

v) (Associative property on scalars) $(st)u = s(tu)$;

vi) (Distributive properties on vectors) $s(u + v) = su + sv$;

vii) (Distributive properties on scalars) $(s + t)u = su + tu$;

viii) (Identity element of scalar multiplication) $1u = u$.

Notice that for different sets we have different definitions of addition and scalar multiplication. The properties we described for $\mathbb{R}^n$ are not anymore for free for every sets of objects.

Example 3.1.1. If we consider the set S of all 2×2 invertible matrices, we cannot expect that addition of two invertible matrices is again invertible; for instance,

$$A = \begin{bmatrix} 1 & 0 \\ 0 & 1 \end{bmatrix}, B = \begin{bmatrix} 0 & 1 \\ 1 & 0 \end{bmatrix},$$

both are invertible, but $A + B$ is not. That is, invertibility is not preserved under addition and the set S does not satisfy the closedness property. □

However, we do have many occasions that a common property is automatically preserved under innocent operations without the need to have additional verifications, and we expect the same easiness as in $\mathbb{R}^n$ when working on different sets of objects equipped with their own operations of addition and scalar multiplication. We can check the following sets with the questions we asked for $\mathbb{R}^n$:

Example 3.1.2.

i) The set of all $m \times n$ matrices;

ii) The set of all $m \times m$ symmetric matrices;

iii) The set of all $m \times m$ skew-symmetric matrices A with $A^T = -A$;

iv) The set of all solutions of $Ax = 0$ in $\mathbb{R}^n$;

v) The set of all linear combinations of vectors in $\{(1, 1, 0), (1, 0, 0)\} \subset \mathbb{R}^3$

vi) The set with addition identity only: $\{\mathbf{0}\}$;

vii) The set of all polynomials;

viii) The set of all convergent sequences in $\mathbb{R}$;

ix) The set of all continuous real functions;

x) The set of all differentiable real functions;

xi) The set of all Riemann integrable real functions on $[a, b]$. □

Definition 3.1.3. Let V be a set equipped with addition $+$ and scalar multiplication over a scalar field K (we assume it is either $\mathbb{R}$ or $\mathbb{C}$). If V is closed under addition and scalar multiplication, and the following conditions (A1)–(A8) are satisfied, we call V a vector space (or linear space) over the field K. If the scalar field K is $\mathbb{R}$, we call V a real vector space. If $K = \mathbb{C}$, we call V a complex vector space. We discuss real vector space by default.

A1) For every $x, y \in V$, $x + y = y + x$.

A2) For every $x, y, z \in V$, we have $(x + y) + z = x + (y + z)$.

A3) There exists a $0 \in V$ such that $x + 0 = x = 0 + x$ for every $x \in V$.

A4) For every $x \in V$, there exists $w \in V$ such that $x + w = 0$.

A5) For every $x, y \in V$ and $k \in K$, $k(x + y) = kx + ky$.

A6) For every $x \in V$ and $k, t \in K$, $(k + t)x = kx + tx$.

A7) For every $x \in V$ and $k, t \in K$, $k(tx) = (kt)x$.

A8) For every $x \in V$, $1x = x$.

We know that $\mathbb{R}^n$ has many subspaces. We also have the notion of **subspace** of general vector space.

> **Definition 3.1.4.** Let V be a vector space. A subset $E \subset V$ is called a subspace if it is closed under addition and scalar multiplication. That is, for every x, $y \in E$, k a scalar,
>
> $$x + y \in E, \quad kx \in E.$$

We can verify that

> **Lemma 3.1.5.** If E is a subspace of a vector space V, then (A1)–(A8) are also satisfied by E and E itself is a vector space.

Example 3.1.6.

Let $M_{n \times n}$ denote the vector space of all $n \times n$ matrices. Then the set of all $n \times n$ symmetric matrices is a subspace of $M_{n \times n}$.

Let V be a vector space and $v_0 \in V$. Then $\{kv_0 : k \in \mathbb{R}\}$ is a subspace of V.

For every vector space V, the set $\{\mathbf{0}\}$ is a (trivial) subspace of V. $\qquad\square$

Example 3.1.7. Let V be a vector space and $\{u, v\} \subset V$. Let E be the set of all linear combinations of u, v. Then E is a subspace of V. Indeed, we need only to check the closure property for S.

<u>Closed under addition:</u> For every x, $y \in S$, there exist scalars c_1, c_2, c_1', c_2' such that $x = c_1 u + c_2 v$, $y = c_1' u + c_2' v$. Then we have

$$x + y = (c_1 + c_1')u + (c_2 + c_2')v,$$

which is again a linear combination of u, v. We have $x + y \in S$.

<u>Closed under scalar multiplication:</u> For every $x \in S$, there exist scalars c_1, c_2 such that $x = c_1 u + c_2 v$. Then for every scalar t we have

$$tx = t(c_1 u + c_2 v) = tc_1 u + tc_2 v,$$

which is also a linear combination of u, v. We have $tx \in S$. By definition of subspaces, S is a subspace of V. $\qquad\square$

We call the set of all linear combinations of vectors from a given subset $S \subset V$ the **span** of S, denoted by $\text{span}(S)$. Note that a linear combination of vectors is a linear combination of **finitely many** vectors. By the same token of Example 3.1.7, we can show that

> **Lemma 3.1.8.** If S is a subset of the vector space V, then $\text{span}(S)$ is a subspace of V.

We notice that if we add a linear combination of the vectors from a spanning set, say $S = \{u, v\}$, it will not change the span of S. For instance,

$$\text{span}\{u, v\} = \text{span}\{u, v, u + v\}.$$

If a spanning set S is linearly independent, it has the minimal number of vectors to span the vector space $\text{span}(S)$. At the same time, S also has the maximal number of linearly independent vectors in $\text{span}(S)$ because any additional one from $\text{span}(S)$ will create redundancy in S. As such, we have the following definition,

Definition 3.1.9. If S is linearly independent, we call the number of vectors in S the **dimension** of the vector space $V = \text{span}(S)$, denoted $\dim V$, and call S a **basis** of the vector space $V = \text{span}(S)$.

Notice in the definition, a set of vectors S qualifies for a basis of a vector space V, if and only if,

1) S is linearly independent in V;
2) S spans V.

Moreover, we notice that the dimension of a vector space is independent of a specific basis. Indeed we have the following **dimension theorem**,

Theorem 3.1.10. All bases for a vector space have the same number of vectors.

Proof. We consider bases with finitely many vectors. Suppose the vector space V has two bases $\{v_1, v_2, \cdots, v_m\}$ and $\{w_1, w_2, \cdots, w_n\}$ with $m > n$. Then there exists an $m \times n$ matrix $A = (a_{ij})$ such that

$$[v_1, v_2, \cdots, v_m] = [w_1, w_2, \cdots, w_m]A.$$

Then $Ax = 0$ has at least one nontrivial solution $x = x_0$ since the reduced row echelon form of A will always have zero row. Then

$$[v_1, v_2, \cdots, v_m]x_0 = [w_1, w_2, \cdots, w_m]Ax_0 = 0.$$

That is, there exists a nontrivial linear combination of $\{v_1, v_2, \cdots, v_m\}$ that equals zero vector. $\{v_1, v_2, \cdots, v_m\}$ is linearly dependent. This is a contradiction.   □

Example 3.1.11. Let $S = \{u, v\}$ be linearly independent vectors. Justify that $\{u + v, u - v\}$ is a basis of $\text{span}(S)$ and find the dimension of $\text{span}(S)$.

Proof. We show that the set of vectors $\{u + v,\ u - v\}$ is linearly independent and spans the vector space span(S). Consider the vector equation:

$$c_1(u + v) + c_2(u - v) = 0,$$

and we have $(c_1 + c_2)u + (c_1 - c_2)v = 0$. Since u, v are linearly independent, we have

$$\begin{cases} c_1 + c_2 = 0 \\ c_1 - c_2 = 0 \end{cases} \Rightarrow \begin{cases} c_1 = 0 \\ c_2 = 0, \end{cases}$$

$\{u + v,\ u - v\}$ is linearly independent and is a basis of span $\{u + v,\ u - v\}$.
Since $\{u + v,\ u - v\} \subset$ span $\{u,\ v\}$, we have

$$\text{span}\,\{u + v,\ u - v\} \subset \text{span}\,\{u,\ v\}.$$

Notice that $u = \frac{u+v}{2} + \frac{u-v}{2}, v = \frac{u+v}{2} - \frac{u-v}{2}$. We have

$$\text{span}\,\{u,\ v\} \subset \text{span}\,\{u + v,\ u - v\},$$

and hence

$$\text{span}\,\{u,\ v\} = \text{span}\,\{u + v,\ u - v\}.$$

Therefore, $\{u+v,\ u-v\}$ is also a basis of span(S). The dimension of span $\{u + v,\ u - v\}$ is 2.

$\square$

We close this section with an example on how to find the span of vectors in $\mathbb{R}^3$.

Example 3.1.12. Find the equation of the plane in $\mathbb{R}^3$ spanned by $v_1 = (1,\ 1,\ 0)$ and $v_2 = (0,\ 1,\ 1)$.

Solution: Let S denote the plane. For every $(x,\ y,\ z) \in S$, it is a linear combination of v_1 and v_2. That is, the vector equation

$$c_1 v_1 + c_2 v_2 = (x,\ y,\ z)$$

is always consistent for c_1 and c_2. Applying Gaussian elimination to the augmented matrix, we have

$$\begin{bmatrix} 1 & 0 & x \\ 1 & 1 & y \\ 0 & 1 & z \end{bmatrix} \xrightarrow{R_2 - R_1} \begin{bmatrix} 1 & 0 & x \\ 0 & 1 & y - x \\ 0 & 0 & z \end{bmatrix} \xrightarrow{R_3 - R_2} \begin{bmatrix} 1 & 0 & x \\ 0 & 1 & y - x \\ 0 & 0 & z - (y - x) \end{bmatrix}.$$

Then we have $x - y + z = 0$, which is the equation of the plane.

$\square$

Exercise 3.1.13.

1. Which of the following subsets of $\mathbb{R}^3$ are subspaces of $\mathbb{R}^3$? If yes, find a basis and the dimension of each of the subspaces.

   i) $\{(x, y, z) : x + y + z = 0\}$;

   ii) $\{(x, y, z) : xyz = 0\}$;

   iii) $\{(x, y, z) : x + y + z = 1\}$;

   iv) $\{(x, y, z) : x = y = z\}$;

   v) $\{(x, y, z) : y = z\}$.

2. Let V be a vector space. Show that the zero vector $\mathbf{0}$ is unique.

3. Let V be a vector space. For every $x \in V$, the negative w such that $x + w = \mathbf{0}$ is unique.

4. Let V be a vector space. For every $x \in V$, $0x = \mathbf{0}$.

5. Find a basis and the dimension of the $\{A \in M_{22} : A^T = -A\}$, where M_{22} denotes the vector space of all 2×2 matrices.

6. Find a basis and the dimension of the $\{A \in M_{33} : A^T = -A\}$, where M_{33} denotes the vector space of all 3×3 matrices.

7. Show that $S = \{(1, 1), (1, 0)\}$ is a basis of $\mathbb{R}^2$.

8. Let A be an $n \times n$ matrix. Show that $V = \{B : AB = BA\}$ is a subspace of M_{nn}.

9. Let V be a vector space. U and W are subspaces of V. Show that $U \cap W$ is a subspace of V.

10. Give an example to show that the union of two subspaces may not be a subspace.

11. Let V be a vector space. U and W are subspaces of V. Define $U + W$ by

$$U + W = \{x + y : x \in U, y \in W\}.$$

Show that $U + W$ is a subspace of V.

12. Let u and v be linearly independent vectors in $\mathbb{R}^2$. Show that $\mathbb{R}^2 = \text{span}\{u, v\}$.

13. Show that if the subset S has m vectors in the n-dimensional space V with $m > n$, then S must be linearly dependent. (One may use the proof method for Theorem 3.1.10.)

14. Find the equation of the plane in $\mathbb{R}^3$ spanned by $v_1 = (-1, 1, 0)$ and $v_2 = (0, 1, -1)$.

15. We call an $(n-1)$-dimensional subspace of an n-dimensional vector space V a **hyperplane** in V. Find the equation of the hyperplane in $\mathbb{R}^4$ spanned by $v_1 = (-1, 1, 0, 0)$, $v_2 = (0, -1, 1, 0)$ and $v_3 = (0, 0, -1, 1)$.

3.2 Nullspace, row space and column space

Let A be an $m \times n$ matrix. The set of all solutions $N(A) = \{x \in \mathbb{R}^n : Ax = 0\}$ is a subspace of $\mathbb{R}^n$. Indeed, by linearity of matrix multiplication, for every $u, v \in N(A)$ and for every scalar t, we have

$$A(u + v) = Au + Av = 0 + 0 = 0,$$
$$A(tu) = tAu = t0 = 0,$$

which imply that $u + v \in N(A)$ and $tu \in N(A)$.

Definition 3.2.1. Let A be an $m \times n$ matrix. The set of all vectors

$$N(A) = \{x \in \mathbb{R}^n : Ax = 0\}$$

is called the **nullspace** of A.

We can obtain a nullspace by means of elimination. Indeed, if E is a product of elementary matrices, then

$$EAx = 0 \Leftrightarrow Ax = 0.$$

Therefore, the nullspace of A is not changed after elimination. We can find the nullspace using the reduced row echelon form.

Example 3.2.2. Let $E = \begin{bmatrix} 1 & 0 & 0 & 1 \\ 0 & 0 & 1 & 0 \end{bmatrix}$. Then in the solution of $Ex = 0$ with $x = (x_1, x_2, x_3, x_4)$, x_1, x_3 are leading variables and x_2, x_4 are free variables.

$$\begin{bmatrix} x_1 \\ x_2 \\ x_3 \\ x_4 \end{bmatrix} = \begin{bmatrix} -x_4 \\ x_2 \\ 0 \\ x_4 \end{bmatrix} = \begin{bmatrix} 0 \\ x_2 \\ 0 \\ 0 \end{bmatrix} + \begin{bmatrix} -x_4 \\ 0 \\ 0 \\ x_4 \end{bmatrix} = x_2 \begin{bmatrix} 0 \\ 1 \\ 0 \\ 0 \end{bmatrix} + x_4 \begin{bmatrix} -1 \\ 0 \\ 0 \\ 1 \end{bmatrix}.$$

Therefore, we have

$$N(E) = \text{span} \left(\begin{bmatrix} 0 \\ 1 \\ 0 \\ 0 \end{bmatrix}, \begin{bmatrix} -1 \\ 0 \\ 0 \\ 1 \end{bmatrix} \right),$$

and $\dim N(E) = 2$. We call $\begin{bmatrix} 0 \\ 1 \\ 0 \\ 0 \end{bmatrix}$ and $\begin{bmatrix} -1 \\ 0 \\ 0 \\ 1 \end{bmatrix}$ special solutions of $Ex = 0$, which constitute a basis of $N(E)$. The number of free variables is the dimension of the nullspace. $\square$

We know an $m \times n$ matrix A is a rectangular array of numbers. Every row of A can be regarded as a vector in $\mathbb{R}^n$ and we call it a row vector. Every column of A can be regarded as a vector in $\mathbb{R}^m$ and we call it a column vector.

Definition 3.2.3. Let A be an $m \times n$ matrix. The span of all rows of A is a subspace of $\mathbb{R}^n$, and is called the **row space** of A denoted by $R(A)$. The span of all columns of A is a subspace of $\mathbb{R}^m$, and is called the **column space** of A denoted by $C(A)$.

Example 3.2.4. Let $A = \begin{bmatrix} 1 & 0 & 0 \\ 0 & 1 & 0 \end{bmatrix}$. Then we have the row space

$$
\begin{aligned}
R(A) &= \mathrm{span}\{(1,\,0,\,0),(0,\,1,\,0)\} \\
&= \{u \in \mathbb{R}^3 : u = x(1,\,0,\,0) + y(0,\,1,\,0),\ x,\ y \in \mathbb{R}\} \\
&= \{u \in \mathbb{R}^3 : u = (x,\,y,\,0),\ x,\ y \in \mathbb{R}\} \\
&= \text{the } xy\text{-plane in } \mathbb{R}^3,
\end{aligned}
$$

and the column space

$$
\begin{aligned}
C(A) &= \mathrm{span}\{(1,\,0),(0,\,1)\} \\
&= \{v \in \mathbb{R}^2 : v = x(1,\,0) + y(0,\,1),\ x,\ y \in \mathbb{R}\} \\
&= \{v \in \mathbb{R}^2 : v = (x,\,y),\ x,\ y \in \mathbb{R}\} \\
&= \mathbb{R}^2.
\end{aligned}
$$

$\square$

We have seen in Example 3.2.4 that the row space can be obtained when the matrix is in reduced row echelon form (See Example 2.1.3 for the definition of reduced row echelon form.) To be specific,

---

If a matrix R is in reduced row echelon form, the row space of R is spanned by the row vectors with pivots.

If a matrix is not in reduced row echelon form, we can reduce it into reduced row echelon form whose basis for row space is the pivot rows. The question is: will the row space be the same after elementary row operations? To find out an answer, let us assume we have an $m \times n$ matrix A, which has been reduced into B by a left multiplication of the elementary matrix E. Then we have

$$A = EB \quad \text{and} \quad B = E^{-1}A.$$

By matrix multiplication, we have

$$\text{Row } i \text{ of } A = (\text{Row } i \text{ of } E)B$$

$$= (\text{Row } i \text{ of } E) \begin{bmatrix} r_1 \\ r_2 \\ \vdots \\ r_m \end{bmatrix}$$

$$= (\text{a linear combination of the rows of } B).$$

Therefore, every row of A is a linear combination of the rows of B. That is, $R(A) \subset R(B)$. By the same token on $B = E^{-1}A$, we have $R(B) \subset R(A)$. Therefore, $R(A) = R(B)$.

> **Lemma 3.2.5.** The row space of A is not changed after elementary row operations.

Unfortunately elementary row operations may change the column space of a matrix. For example,

$$A = \begin{bmatrix} 1 & 2 \\ 2 & 4 \end{bmatrix}, \quad E = \begin{bmatrix} 1 & 0 \\ -2 & 1 \end{bmatrix}, \quad EA = \begin{bmatrix} 1 & 2 \\ 0 & 0 \end{bmatrix}.$$

The column space of A is span$\{(1, 2)\}$. But after the row operation, the row space of EA becomes span$\{(1, 0)\}$, which is not equal to span$\{(1, 2)\}$. However, this inequality does not mean row operations are useless for finding column space. Indeed, a row operation E on A does not change linear dependency among the columns of A, noticing that $Ax = 0 \Leftrightarrow EAx = 0$, i.e.,

$$Ax = [c_1 : c_2 : \cdots : c_n] \begin{bmatrix} x_1 \\ x_2 \\ \vdots \\ x_n \end{bmatrix} = 0 \Leftrightarrow EAx = 0.$$

If the only solution of $Ax = 0$ is the trivial solution $x = (x_1, x_2, \cdots, x_n) = 0$, then all columns of A are linearly independent; so are the columns of EA.

If $Ax = 0$ has the nontrivial solution $\mathbf{x} = (x_1, x_2, \cdots, x_n) \neq 0$, say, $\mathbf{x} = (1, \frac{1}{2}, 1, 0, \cdots, 0)$, then we have the following linear dependency among columns of A:

$$1 \cdot (\text{Column 1 of } A) + \frac{1}{2} \cdot (\text{Column 2 of } A) + 1 \cdot (\text{Column 3 of } A) = 0.$$

Since $\mathbf{x}$ is also a solution of $EAx = 0$, we have

$$1 \cdot (\text{Column 1 of } EA) + \frac{1}{2} \cdot (\text{Column 2 of } EA) + 1 \cdot (\text{Column 3 of } EA) = 0,$$

which is the same set of linear combinations for the **corresponding columns** of A.

Lemma 3.2.6. The linear dependency of columns of A is not changed after elementary row operations.

Example 3.2.7. Consider the 4×5 matrix $A = [\alpha_1 \mid \alpha_2 \mid \alpha_3 \mid \alpha_4 \mid \alpha_5]$, where the columns are

$$\alpha_1 = \begin{bmatrix} 1 \\ -1 \\ 0 \\ 1 \end{bmatrix}, \ \alpha_2 = \begin{bmatrix} -2 \\ 3 \\ 1 \\ 2 \end{bmatrix}, \ \alpha_3 = \begin{bmatrix} 1 \\ 0 \\ 1 \\ 5 \end{bmatrix}, \ \alpha_4 = \begin{bmatrix} 1 \\ 2 \\ 3 \\ 13 \end{bmatrix}, \ \alpha_5 = \begin{bmatrix} 2 \\ -2 \\ 4 \\ 5 \end{bmatrix}.$$

a) Find a set of vectors in $\{\alpha_1, \alpha_2, \alpha_3, \alpha_4, \alpha_5\}$ which is a basis of the column space of A.

b) Find the dimension of the column space of A.

c) Given the basis of the column space of A which has been obtained in part a), write the nonbasis vectors in $\{\alpha_1, \alpha_2, \alpha_3, \alpha_4, \alpha_5\}$ as linear combinations of the basis vectors.

   Solution: a) By the elementary row operations on A we have

$$\begin{bmatrix} 1 & -2 & 1 & 1 & 2 \\ -1 & 3 & 0 & 2 & -2 \\ 0 & 1 & 1 & 3 & 4 \\ 1 & 2 & 5 & 13 & 5 \end{bmatrix} \xrightarrow[R_4 - R_1]{R_2 + R_1} \begin{bmatrix} 1 & -2 & 1 & 1 & 2 \\ 0 & 1 & 1 & 3 & 0 \\ 0 & 1 & 1 & 3 & 4 \\ 0 & 4 & 4 & 12 & 3 \end{bmatrix}$$

$$\xrightarrow[R_4 - 4R_2]{R_1 + 2R_2; \ R_3 - R_2} \begin{bmatrix} 1 & 0 & 3 & 7 & 2 \\ 0 & 1 & 1 & 3 & 0 \\ 0 & 0 & 0 & 0 & 4 \\ 0 & 0 & 0 & 0 & 3 \end{bmatrix}$$

$$\xrightarrow{R_3/(4)} \begin{bmatrix} 1 & 0 & 3 & 7 & 2 \\ 0 & 1 & 1 & 3 & 0 \\ 0 & 0 & 0 & 0 & 1 \\ 0 & 0 & 0 & 0 & 3 \end{bmatrix}$$

$$\xrightarrow[R_4 - 3R_3]{R_1 - 2R_3} \begin{bmatrix} 1 & 0 & 3 & 7 & 0 \\ 0 & 1 & 1 & 3 & 0 \\ 0 & 0 & 0 & 0 & 1 \\ 0 & 0 & 0 & 0 & 0 \end{bmatrix}.$$

We note that the leading 1's are in columns 1, 2 and 5 of the reduced row echelon form of A. Then, correspondingly, $\{\alpha_1, \alpha_2, \alpha_5\}$ is a basis of the column space of A.

   b) Since $\{\alpha_1, \alpha_2, \alpha_5\}$ is a basis of the column space of A, the dimension of the column space of A is 3.

c) Let c_1, c_2, c_3, c_4, c_5 be the columns of the reduced row echelon form of A which is derived in part a). Then we have

$$c_3 = 3c_1 + c_2, \ c_4 = 7c_1 + 3c_2.$$

Since elementary row operations do not change linear dependence of the columns of A, we have

$$\alpha_3 = 3\alpha_1 + \alpha_2, \ \alpha_4 = 7\alpha_1 + 3\alpha_2.$$

$\square$

Exercise 3.2.8.

1. Let

$$A = \begin{bmatrix} 1 & 4 & 9 & 1 \\ 2 & 5 & 1 & -1 \\ 3 & 7 & 0 & 3 \end{bmatrix}.$$

i) Find the columns of A which constitute a basis of its column space; ii) Write the nonbasis columns of A into linear combinations of the basis columns; iii) Find the nullspace of A and determine a basis.

2. Find the rows of A (not its reduced row echelon form) which constitute a basis of its row space, where

$$A = \begin{bmatrix} 1 & 2 & 3 \\ 4 & 5 & 7 \\ 9 & 1 & 0 \\ 1 & -1 & 3 \end{bmatrix}.$$

3. Show that $N(A) = N(EA)$ if E is invertible.

4. Construct an example of a matrix A and an elementary matrix E such that

$$C(A) \neq C(EA).$$

5. Let $f : \mathbb{R}^n \to \mathbb{R}^m$ be a function defined by $f(x) = Ax$ where A is an $m \times n$ real matrix. i) Show that the range of f is the column space of A. ii) Show that f is a linear function. (See Exercise 2.1.6 for the definition of linearity.)

6. Let A be an $m \times n$ real matrix. Find a linear function $g : \mathbb{R}^m \to \mathbb{R}^n$ such that the range of g is the row space of A.

7. Let $f : \mathbb{R}^n \to \mathbb{R}^m$ be a linear function. Show that there exists a unique $m \times n$ real matrix A such that $f(x) = Ax$.

8. Let $f : \mathbb{R}^n \to \mathbb{R}^m$ and $g : \mathbb{R}^m \to \mathbb{R}^r$ be a linear function. i) Show that $g \circ f : \mathbb{R}^n \to \mathbb{R}^r$ is also a linear function; ii) Find a matrix C such that $(g \circ f)(x) = Cx$ for every $x \in \mathbb{R}^n$.

9. Let V be a real vector space and $u, v \in V$. Show that

$$\text{span}\{u, v\} = \text{span}\{3u + 2v, 4u - 5v\}.$$

10. Let V and W be real vector spaces and A a subset of V. Let $f(A)$ be the image of A under the function $f : V \to W$. That is, $f(A) = \{f(x) : x \in A\}$. Show that if f is linear then $f(\text{span}(A)) = \text{span}(f(A))$.

3.3   Solutions of $Ax = b$

Now we turn to discuss the solution structure of $Ax = b$. We examine the following example:

Example 3.3.1. Let $E = \begin{bmatrix} 1 & 0 & 0 & 1 \\ 0 & 0 & 1 & 0 \end{bmatrix}$ and consider $Ex = b$ with $b = \begin{bmatrix} b_1 \\ b_2 \end{bmatrix}$. We have the augmented matrix

$$[E : b] = \begin{bmatrix} 1 & 0 & 0 & 1 & b_1 \\ 0 & 0 & 1 & 0 & b_2 \end{bmatrix}.$$

Then in the solution $x = (x_1, x_2, x_3, x_4)$, x_1, x_3 are still pivot (leading) variables and x_2, x_4 are free variables.

$$\begin{bmatrix} x_1 \\ x_2 \\ x_3 \\ x_4 \end{bmatrix} = \begin{bmatrix} b_1 - x_4 \\ x_2 \\ b_2 \\ x_4 \end{bmatrix} = \begin{bmatrix} b_1 \\ 0 \\ b_2 \\ 0 \end{bmatrix} + \begin{bmatrix} 0 \\ x_2 \\ 0 \\ 0 \end{bmatrix} + \begin{bmatrix} -x_4 \\ 0 \\ 0 \\ x_4 \end{bmatrix} = \begin{bmatrix} b_1 \\ 0 \\ b_2 \\ 0 \end{bmatrix} + x_2 \begin{bmatrix} 0 \\ 1 \\ 0 \\ 0 \end{bmatrix} + x_4 \begin{bmatrix} -1 \\ 0 \\ 0 \\ 1 \end{bmatrix}.$$

We notice that $x_p = (b_1, 0, b_2, 0)$ is a particular solution when we set the free variables zero, and $x_n = x_2(0, 1, 0, 0) + x_4(-1, 0, 0, 1)$. Namely the solution can be written $x = x_p + x_n$. □

It is not by chance we have the phenomenon in Example 3.3.1 that a solution of $Ax = b$ is the sum of its particular solution x_p and a general solution x_n of the homogeneous system $Ax = 0$.

> **Theorem 3.3.2.** x is a solution of $Ax = b$ if and only if $x = x_p + x_n$, where x_p is a particular solution of $Ax = b$ and x_n a solution of the homogeneous system $Ax = 0$.

Proof. "$\Leftarrow$" If $x = x_p + x_n$, we have $Ax = A(x_p + x_n) = Ax_p + Ax_n = b + 0 = b$. That is $x = x_p + x_n$ is a solution.

"$\Rightarrow$" If x is a solution of $Ax = b$, then for every particular solution x_p, we have $A(x - x_p) = Ax - Ax_p = b - b = 0$. That is, $x_n = x - x_p$ is a solution of the homogeneous system $Ax = 0$. □

By Theorem 3.3.2, we know that every two solutions of $Ax = b$ differ by a solution of the homogeneous system $Ax = 0$. Hence the general solution of $Ax = b$ is a particular solution of $Ax = b$ plus the general solution of $Ax = 0$.

There is one more question remaining: when is $Ax = b$ solvable? Recall that Ax is just a linear combination of the columns of A, we have

Theorem 3.3.3. $Ax = b$ is solvable if and only if $b \in C(A)$.

In terms of augmented matrix, $Ax = b$ is solvable if and only if the last column of the reduced row echelon form of $[A : b]$ is not a pivot column. We rephrase Theorem 3.3.3 as

Theorem 3.3.4. $Ax = b$ is solvable if and only if $rank(A) = rank([A : b])$, where $rank(A)$ is the dimension of the column space of A, which equals the number of pivots.

Example 3.3.5. (Example 2.1.5 revisited.) Let k be a real number. Consider the following linear system of equations:

$$\begin{cases} x_2 + x_3 + x_4 = -1 \\ 2x_1 + x_2 + 3x_3 + x_4 = -2 \\ 2x_1 - 2x_2 + 4x_3 + 2x_4 = -3 \\ 3x_2 - x_3 - x_4 = k. \end{cases} \tag{3.1}$$

Find all possible values of k such that system (3.1) has a unique solution, has no solutions and has infinitely many solutions.

Solution: We re-write the system of linear equations in the matrix form $Au = b$, where

$$A = \begin{bmatrix} 0 & 1 & 1 & 1 \\ 2 & 1 & 3 & 1 \\ 2 & -2 & 4 & 2 \\ 0 & 3 & -1 & -1 \end{bmatrix}, \quad u = \begin{bmatrix} x_1 \\ x_2 \\ x_3 \\ x_4 \end{bmatrix} \quad \text{and} \quad b = \begin{bmatrix} -1 \\ -2 \\ -3 \\ k \end{bmatrix}.$$

Then the corresponding augmented matrix is

$$[A : b] = \begin{bmatrix} 0 & 1 & 1 & 1 & -1 \\ 2 & 1 & 3 & 1 & -2 \\ 2 & -2 & 4 & 2 & -3 \\ 0 & 3 & -1 & -1 & k \end{bmatrix}.$$

By the elementary row operations on $[A : b]$ we have

$$\begin{bmatrix} 0 & 1 & 1 & 1 & -1 \\ 2 & 1 & 3 & 1 & -2 \\ 2 & -2 & 4 & 2 & -3 \\ 0 & 3 & -1 & -1 & k \end{bmatrix} \xRightarrow{\text{eliminations}} \begin{bmatrix} 1 & 0 & 0 & -1 & \frac{1}{2} \\ 0 & 1 & 0 & 0 & 0 \\ 0 & 0 & 1 & 1 & -1 \\ 0 & 0 & 0 & 0 & k-1 \end{bmatrix}$$

from which we can tell that $\operatorname{rank}(A) = 3$, but $\operatorname{rank}([A : b])$ depends on the value of k.

i) If $k \neq 1$, $\operatorname{rank}([A : b]) = 4 \neq \operatorname{rank}(A)$. By Theorem 3.3.4, system $Ax = b$ has no solution. From the point of view of Theorem 3.3.3, system $Ax = b$ has no solution since $b \notin C(A)$. Recall that elementary row operations do not change linear dependency among the columns of a matrix.

ii) If $k = 1$, $\operatorname{rank}([A : b]) = 3 = \operatorname{rank}(A)$. By Theorem 3.3.4, system $Ax = b$ has at least one solution. Note that the homogeneous system $Ax = 0$ has a nontrivial nullspace because $\operatorname{rank}(A) = 3$ is less than the column number 4 of A and there is a free variable for the solution. Hence system $Ax = b$ has infinitely many solutions. Let $x_4 = t$, where t is an arbitrary real number. The solution is

$$\begin{bmatrix} x_1 \\ x_2 \\ x_3 \\ x_4 \end{bmatrix} = \begin{bmatrix} \frac{1}{2} + t \\ 0 \\ -1 - t \\ t \end{bmatrix} = \begin{bmatrix} \frac{1}{2} \\ 0 \\ -1 \\ 0 \end{bmatrix} + t \begin{bmatrix} 1 \\ 0 \\ -1 \\ 1 \end{bmatrix}, \ t \in \mathbb{R},$$

where $x_p = \left(\frac{1}{2}, 0, -1, 0\right)$ is a particular solution of $Ax = b$ and $x_n = t\left(1, 0, -1, 1\right), t \in \mathbb{R}$ is the general solution of the homogeneous system $Ax = 0$.

□

Exercise 3.3.6.

1. Let the following matrices be the augmented matrices $[A : b]$ of the system $Ax = b$. i) Determine whether the system is consistent or not. ii) Find all possible solutions if $Ax = b$ is consistent. iii) If $Ax = b$ is consistent, write b into a linear combination of the columns of A.

$$a) \ \begin{bmatrix} 1 & 3 & 5 & 7 \\ 3 & 0 & 2 & 6 \end{bmatrix}, \quad b) \ \begin{bmatrix} 1 & 3 & 5 & 7 \\ 3 & 0 & 2 & 6 \\ 0 & 1 & 2 & 5 \end{bmatrix}, \quad c) \ \begin{bmatrix} 1 & 3 & 6 & 7 \\ 3 & 0 & 0 & 6 \\ 0 & 1 & 2 & 5 \end{bmatrix}.$$

2. Let A be an $n \times n$ matrix. Show that $Ax = b$ has a unique solution if and only if A is invertible.

3. Let

$$A = \begin{bmatrix} 1 & 3 & 5 & 7 \\ 3 & 0 & 2 & 6 \\ 0 & 1 & 2 & 5 \\ 3 & 0 & 3 & 12 \end{bmatrix}, \ b = \begin{bmatrix} 1 \\ 2 \\ 3 \\ k \end{bmatrix}.$$

Find conditions on $k \in \mathbb{R}$ such that $Ax = b, x \in \mathbb{R}^n$ has 1) a unique solution; 2) no solution; 3) infinitely many solutions, respectively.

4. Let A be an $m \times n$ matrix. Show that if $Ax = b$ has two distinct solutions, then it has infinitely many solutions.

5. Show that if x_1, x_2 are both solutions of $Ax = b$, then i) $x_1 - x_2$ is a solution of $Ax = 0$; ii) for every $t \in \mathbb{R}$, $x_1 + t(x_1 - x_2)$ is a solution of $Ax = b$.

6. Let A and B be $n \times n$ real matrices and $x_0 \in \mathbb{R}^n$ a solution of $Ax = b$. Show that i) x_0 is also a solution of $BAx = Bb$; ii) a solution of $BAx = Bb$ may not be a solution of $Ax = b$.

3.4 Rank of matrices

By Lemma 3.2.5 and Lemma 3.2.6, for a given $m \times n$ matrix, the dimension of the row space is the number of leading 1's (or pivot 1's) in the reduced row echelon form. The dimension of the column space is the number of leading 1's, too. Therefore, we have

> **Lemma 3.4.1.** Let A be an $m \times n$ matrix. The dimensions of the row space and column space of A are equal.

We call the dimension of the row space of a matrix A the **rank** of A, which is also equal to the dimension of the column space. We call the dimension of the nullspace of A the **nullity** of A.

Example 3.4.2. Let

$$A = \begin{bmatrix} 1 & 3 & 5 & 7 \\ 3 & 0 & 2 & 6 \\ 0 & 1 & 2 & 5 \end{bmatrix}.$$

To find the rank of A, we use elementary row operations to reduce A into row echelon form:

$$\begin{bmatrix} 1 & 3 & 5 & 7 \\ 3 & 0 & 2 & 6 \\ 0 & 1 & 2 & 5 \end{bmatrix} \xrightarrow{R_2 - 3R_1} \begin{bmatrix} 1 & 3 & 5 & 7 \\ 0 & -9 & -13 & -15 \\ 0 & 1 & 2 & 5 \end{bmatrix}$$

$$\xrightarrow{R_2 + 9R_3} \begin{bmatrix} 1 & 3 & 5 & 7 \\ 0 & 0 & 5 & 30 \\ 0 & 1 & 2 & 5 \end{bmatrix}$$

$$\xrightarrow{R_3 \leftrightarrow R_2} \begin{bmatrix} 1 & 3 & 5 & 7 \\ 0 & 1 & 2 & 5 \\ 0 & 0 & 5 & 30 \end{bmatrix}$$

$$\xrightarrow{R_3/5} \begin{bmatrix} 1 & 3 & 5 & 7 \\ 0 & 1 & 2 & 5 \\ 0 & 0 & 1 & 6 \end{bmatrix}.$$

Since the row echelon form of A has three pivot 1's, $\text{rank}(A) = 3$. □

Our next question is how to relate $\dim R(A)$, $\dim C(A)$ and $\dim N(A)$. Recall that if the leading 1's in the reduced row echelon form E of A corresponds to leading variables in the solution, the zero rows corresponds to free variables in the solution. Since the number of the leading variables plus the number of free variables is exactly the number of the columns of A, we have the following **counting theorem**,

---

Theorem 3.4.3. Let A be an $m \times n$ matrix. Then we have

$$\dim R(A) + \dim N(A) = n.$$

Remark 3.4.4. Notice that $\dim R(A) \leq m$. If $m < n$, that is, number of columns of A is larger than its number of rows, then $\dim N(A) = n - m \geq 1$. That is, $N(A)$ is nontrivial.

Let A be an $m \times n$ matrix. We call

$$N(A^T) = \{x \in \mathbb{R}^m : A^T x = 0\} = \{x \in \mathbb{R}^m : x^T A = 0\}$$

the **left nullspace** of A. Notice that $\dim R(A) = \dim C(A) = \dim R(A^T)$ because they all are equal to the number of pivots in the reduced row echelon form. Then by Lemma 3.4.3, we have

$$\dim R(A^T) + \dim N(A^T) = m.$$

That is,

Lemma 3.4.5. Let A be an $m \times n$ matrix. Then we have

$$\dim C(A) + \dim N(A^T) = m.$$

Rank one matrix

For matrices with rank one, we can simplify its representation into a product of vectors. Indeed, if $\text{rank}(A) = 1$ where A is $m \times n$, then every row of A is a scalar multiple of a pivot row, say u^T, where by default the vector u is regarded as a column matrix. That is, there exist scalars $c_1, c_2, \cdots, c_m$ such that

$$A = \begin{bmatrix} c_1 u^T \\ c_2 u^T \\ \vdots \\ c_m u^T \end{bmatrix} = \begin{bmatrix} c_1 \\ c_2 \\ \vdots \\ c_m \end{bmatrix} u^T.$$

Let $v = \begin{bmatrix} c_1 \\ c_2 \\ \vdots \\ c_m \end{bmatrix}$. Then $v \neq 0$ and $A = vu^T$, where $u \in \mathbb{R}^m$, $v \in \mathbb{R}^n$. More importantly, we have

> If $\text{rank}(A) = 1$, then $Ax = 0$ is equivalent to $u^T x = 0$, where u^T is a nonzero row of A.

Rank of products

Let A be $m \times n$ and B be $n \times r$. We are interested how $\text{rank}(A), \text{rank}(B)$ and $\text{rank}(AB)$ are related. Indeed, by matrix multiplication:

$$AB = A[b_1 : b_2 : \cdots : b_r] = [Ab_1 : Ab_2 : \cdots : Ab_r],$$

which implies that each column of AB is a linear combination of the columns of A. Therefore we have $\text{rank}(AB) \leq \text{rank}(A)$. Moreover,

$$AB = \begin{bmatrix} a_1 \\ a_2 \\ \vdots \\ a_m \end{bmatrix} B = \begin{bmatrix} a_1 B \\ a_2 B \\ \vdots \\ a_m B \end{bmatrix},$$

which implies that each row of AB is a linear combination of the rows of B. Therefore we have $\text{rank}(AB) \leq \text{rank}(B)$. In summary, we have

> **Theorem 3.4.6.**
>
> $$\text{rank}(AB) \leq \min\{\text{rank}(A), \text{rank}(B)\}.$$

Exercise 3.4.7.

1. Find the ranks and dimensions of the nullspaces of the following matrices .

$$a) \quad \begin{bmatrix} 1 & 3 & 0 & 7 \end{bmatrix}, \quad b) \quad \begin{bmatrix} 0 & 3 & 0 \\ 3 & 9 & 5 \\ 2 & 5 & 5 \end{bmatrix}, \quad c) \quad \begin{bmatrix} 1 & 3 & 0 & 7 \\ 3 & 9 & 5 & 1 \\ 2 & 5 & 5 & 3 \end{bmatrix}, \quad d) \quad \begin{bmatrix} 1 \\ 3 \\ 2 \end{bmatrix}.$$

2. Let A be an $m \times n$ matrix with $m > n$. Show that AA^T is not invertible.

3. Let A be an $m \times n$ matrix. Find all possible vectors x such that $x \in R(A) \cap N(A)$.

4. Let A be an $m \times n$ matrix. If $x_0 \neq 0$ is a solution of $Ax = \mathbf{0}$, then $Ax = \mathbf{0}$ has infinitely many nontrivial solutions.

5. Let A be an $m \times n$ matrix. Show that system $Ax = b$ has a unique solution if and only if

$$rank(A) = rank([A : b]) = n.$$

6. Let A be an $m \times n$ matrix. Show that system $Ax = b$ has infinitely many solutions if and only if

$$rank(A) = rank([A : b]) < n.$$

7. Let

$$A = \begin{bmatrix} 1 & 3 & 5 & 7 \\ 3 & 9 & 15 & 21 \\ 9 & 27 & 45 & 63 \end{bmatrix}.$$

Show that $rank(A) = 1$ and find $u \in \mathbb{R}^3$, $v \in \mathbb{R}^4$ such that $A = vu^T$. Is this decomposition of A unique?

8. Show that if $rank(A) = 1$, then every column of A is a scalar multiple of one specific column of A.

9. Use block multiplication of matrices to show that

$$\text{rank}(A + B) \leq \text{rank}(A) + \text{rank}(B).$$

10. Let $\text{rank}(A) = s$. Find the ranks of the following matrices

$$2A, \quad \begin{bmatrix} A & A \end{bmatrix}, \quad \begin{bmatrix} A \\ A \end{bmatrix}, \quad \begin{bmatrix} A & A \\ A & A \end{bmatrix}.$$

3.5 Bases and dimensions of general vector spaces

Using the notion of span we defined basis and dimension of vector spaces which is spanned by *a priori* known linearly independent set of vectors. We say a vector space is **finite dimensional** if it can be spanned by a finite set of vectors. Otherwise, we say a vector space is **infinite dimensional.** However, so far there is no guarantee that every vector space has a basis. In this section, we explain this issue and show through examples how to find a basis for a given vector space.

Example 3.5.1. Let M_{22} be the set of all 2×2 matrices, which is a vector space with matrix addition and scalar multiplication. For every $A \in M_{22}$, we can represent A as

$$A = \begin{bmatrix} a & b \\ c & d \end{bmatrix}$$

$$= \begin{bmatrix} a & 0 \\ 0 & 0 \end{bmatrix} + \begin{bmatrix} 0 & b \\ 0 & 0 \end{bmatrix} + \begin{bmatrix} 0 & 0 \\ c & 0 \end{bmatrix} + \begin{bmatrix} 0 & 0 \\ 0 & d \end{bmatrix}$$

$$= a \begin{bmatrix} 1 & 0 \\ 0 & 0 \end{bmatrix} + b \begin{bmatrix} 0 & 1 \\ 0 & 0 \end{bmatrix} + c \begin{bmatrix} 0 & 0 \\ 1 & 0 \end{bmatrix} + d \begin{bmatrix} 0 & 0 \\ 0 & 1 \end{bmatrix}.$$

Therefore we have

$$M_{22} = \text{span} \left(\begin{bmatrix} 1 & 0 \\ 0 & 0 \end{bmatrix}, \begin{bmatrix} 0 & 1 \\ 0 & 0 \end{bmatrix}, \begin{bmatrix} 0 & 0 \\ 1 & 0 \end{bmatrix}, \begin{bmatrix} 0 & 0 \\ 0 & 1 \end{bmatrix} \right).$$

Next we show that $\left\{ \begin{bmatrix} 1 & 0 \\ 0 & 0 \end{bmatrix}, \begin{bmatrix} 0 & 1 \\ 0 & 0 \end{bmatrix}, \begin{bmatrix} 0 & 0 \\ 1 & 0 \end{bmatrix}, \begin{bmatrix} 0 & 0 \\ 0 & 1 \end{bmatrix} \right\}$ is linearly independent and consider the vector equation:

$$c_1 \begin{bmatrix} 1 & 0 \\ 0 & 0 \end{bmatrix} + c_2 \begin{bmatrix} 0 & 1 \\ 0 & 0 \end{bmatrix} + c_3 \begin{bmatrix} 0 & 0 \\ 1 & 0 \end{bmatrix} + c_4 \begin{bmatrix} 0 & 0 \\ 0 & 1 \end{bmatrix} = 0,$$

which has only the trivial solution: $(c_1, c_2, c_3, c_4) = (0, 0, 0, 0)$. Then by definition of basis, we have verified that

$$\left\{ \begin{bmatrix} 1 & 0 \\ 0 & 0 \end{bmatrix}, \begin{bmatrix} 0 & 1 \\ 0 & 0 \end{bmatrix}, \begin{bmatrix} 0 & 0 \\ 1 & 0 \end{bmatrix}, \begin{bmatrix} 0 & 0 \\ 0 & 1 \end{bmatrix} \right\}$$

is a basis of M_{22}. □

Example 3.5.2. Let A be an $n \times n$ invertible matrix. Then the set of the columns of A is a basis of $\mathbb{R}^n$, and $C(A) = R(A) = \mathbb{R}^n$. Indeed, the reduced row echelon form U of A is I and hence the set of n-columns of A is linearly independent. If the set of n-columns of A is not a basis, then a basis would have more than n vectors, which is impossible by Theorem 3.1.10 as we know that $\mathbb{R}^n$ has a standard basis $\{e_1, e_2, \cdots, e_n\}$ where

$$e_1 = \begin{pmatrix} 1 \\ 0 \\ \vdots \\ 0 \end{pmatrix}, e_2 \begin{pmatrix} 0 \\ 1 \\ \vdots \\ 0 \end{pmatrix}, \cdots, e_n = \begin{pmatrix} 0 \\ 0 \\ \vdots \\ 1 \end{pmatrix}$$

and the only nonzero entry of e_i is 1 at the i-th coordinate. □

Example 3.5.3. Find a basis of $E = \{x = (x_1, x_2, x_3) \in \mathbb{R}^3 : x_1 - 2x_2 + 3x_3 = 0\}$. E is a plane in $\mathbb{R}^3$ passing through $(0, 0, 0)$. For every $x = (x_1, x_2, x_3) \in E$ we have $x_1 = 2x_2 - 3x_3$ and

$$\begin{bmatrix} x_1 \\ x_2 \\ x_3 \end{bmatrix} = \begin{bmatrix} 2x_2 - 3x_3 \\ x_2 \\ x_3 \end{bmatrix} = \begin{bmatrix} 2x_2 \\ x_2 \\ 0 \end{bmatrix} + \begin{bmatrix} 3x_3 \\ 0 \\ x_3 \end{bmatrix} = x_2 \begin{bmatrix} 2 \\ 1 \\ 0 \end{bmatrix} + x_3 \begin{bmatrix} 3 \\ 0 \\ 1 \end{bmatrix}.$$

That is, $E = \text{span} \left(\begin{bmatrix} 2 \\ 1 \\ 0 \end{bmatrix}, \begin{bmatrix} 3 \\ 0 \\ 1 \end{bmatrix} \right)$. Since $c_1 \begin{bmatrix} 2 \\ 1 \\ 0 \end{bmatrix} + c_2 \begin{bmatrix} 3 \\ 0 \\ 1 \end{bmatrix} = \mathbf{0}$ implies $c_1 = c_2 = $

$0, \left\{ \begin{bmatrix} 2 \\ 1 \\ 0 \end{bmatrix}, \begin{bmatrix} 3 \\ 0 \\ 1 \end{bmatrix} \right\}$ is linearly independent and is a basis for E. We remark that E can be rewritten as

$$
\begin{aligned}
E &= \{x = (x_1, x_2, x_3) \in \mathbb{R}^3 : x_1 - 2x_2 + 3x_3 = 0\} \\
&= \{x = (x_1, x_2, x_3) \in \mathbb{R}^3 : x \cdot (1, -2, 3) = 0\} \\
&= \{x = (x_1, x_2, x_3) \in \mathbb{R}^3 : x \perp (1, -2, 3)\},
\end{aligned}
$$

which means E is the set of all vectors which is orthogonal to the given vector $(1, -2, 3)$ which is called the **normal** of the plane. Indeed, every vector of E is orthogonal to every vector from the subspace $F = \{k(1, -2, 3) : k \in \mathbb{R}\}$. □

Example 3.5.4. Consider the set P_n of all polynomials with degree less than or equal to $n \in \mathbb{R}$. For every $f \in P_n$, we have

$$f(x) = a_0 + a_1 x + a_2 x^2 + \cdots + a_n x^n,$$

which is a linear combination of the set of polynomials $S = \{1, x, \cdots, x^n\}$. That is, $P_n = \mathrm{span}(S)$. We claim that S is a basis for P_n. To check linear independency in S, we consider the vector equation

$$a_0 + a_1 x + a_2 x^2 + \cdots + a_n x^n = 0, \text{ for all } x \in \mathbb{R}.$$

If $(a_0, a_1, \cdots, a_n) \neq 0$, then the vector equation has at most n solutions (instead of all $x \in \mathbb{R}$). Therefore, we have $(a_0, a_1, \cdots, a_n) = 0$ and the set of polynomials $S = \{1, x, \cdots, x^n\}$ is linearly independent. That is, $S = \{1, x, \cdots, x^n\}$ is a basis of P_n, and $\dim P_n = n + 1$. □

Remark 3.5.5. From Example 3.5.4, we notice that if the basis $S = \{1, x, \cdots, x^n\}$ is fixed, every $f \in P_n$ is uniquely determined by the vector consisting of the coefficients $(a_0, a_1, \cdots, a_n) \in \mathbb{R}^{n+1}$. We call $(a_0, a_1, \cdots, a_n) \in \mathbb{R}^{n+1}$ the **coordinate vector** of f with respect to the basis S.

Lemma 3.5.6. Let V be a finite dimensional vector space with a basis $S = \{v_1, v_2, \cdots, v_n\}$. Then the coordinate vector $[x]_S \in \mathbb{R}^n$ of every vector $x \in V$ is unique.

Proof. Suppose not. Then there exists $[x]'_S \in \mathbb{R}^n$ such that

$$x = [v_1, v_2, \cdots, v_n][x]_S = [v_1, v_2, \cdots, v_n][x]'_S,$$

which lead to

$$[v_1, v_2, \cdots, v_n]\left([x]_S - [x]'_S\right) = 0.$$

Since $S = \{v_1, v_2, \cdots, v_n\}$ is a basis and is linearly independent, we have $[x]_S - [x]'_S = 0$ and hence $[x]_S = [x]'_S$. This is a contradiction. □

Change of basis

Let V be a vector space with two bases $B = \{v_1, v_2, \cdots, v_n\}$ and $B' = \{w_1, w_2, \cdots, w_n\}$. Then every v_i, $i = 1, 2, \cdots, n$, is a linear combination of the vectors in B'. That is, there exists an $n \times n$ matrix P such that

$$[v_1, v_2, \cdots, v_n] = [w_1, w_2, \cdots, w_n]P.$$

Then P is invertible because system $Px = 0$ has only the trivial solution. Indeed,

$$
\begin{aligned}
Px = 0 &\Rightarrow [w_1, w_2, \cdots, w_n]Px = 0 \\
&\Rightarrow [v_1, v_2, \cdots, v_n]x = 0 \\
&\Rightarrow x = 0.
\end{aligned}
$$

We call P the **transition matrix** from the basis B to the basis B' and write $P = P_{B \to B'}$.

We are interested how the coordinate vectors are related when a basis is changed into another. Let $x \in V$ be a vector with coordinate vector $[x]_B \in \mathbb{R}^n$ under the basis B, and $[x]_{B'} \in \mathbb{R}^n$ under the basis B'. Then, on the one hand, we have

$$
\begin{aligned}
x &= [v_1, v_2, \cdots, v_n][x]_B \\
&= ([w_1, w_2, \cdots, w_n]P)\,[x]_B \\
&= [w_1, w_2, \cdots, w_n](P[x]_B).
\end{aligned}
$$

On the other hand, we have

$$x = [w_1, w_2, \cdots, w_n][x]_{B'}.$$

Since $B' = \{w_1, w_2, \cdots, w_n\}$ is a basis, $[w_1, w_2, \cdots, w_n](P[x]_B) = [w_1, w_2, \cdots, w_n][x]_{B'}$ implies that

$$P[x]_B = [x]_{B'}.$$

In summary we have

Lemma 3.5.7. Let V be a vector space with two bases $B = \{v_1, v_2, \cdots, v_n\}$ and $B' = \{w_1, w_2, \cdots, w_n\}$. Then there is an $n \times n$ invertible transition matrix $P_{B \to B'}$ such that

$$[v_1, v_2, \cdots, v_n] = [w_1, w_2, \cdots, w_n]P_{B \to B'}$$

and for every $x \in V$ its coordinate vectors with respect to the bases B and B' satisfy

$$P_{B \to B'}[x]_B = [x]_{B'}.$$

For many vector spaces, we have certain bases such as the standard bases with which the coordinate vectors are easy to compute. For example, $\mathbb{R}^n$ has the standard basis $\{e_1, e_2, \cdots, e_n\}$ with which every vector of $\mathbb{R}^n$ coincides with its coordinate vector, and P_n has a standard basis $\{1, x, x^2, \cdots, x^n\}$ with which the coordinate vector of a polynomial is the vector in $\mathbb{R}^{n+1}$ consisting of its coefficients. Moreover, for a general n-dimensional vector space V, each column of the transition matrix $P_{B \to S}$ from the basis $B = \{v_1, v_2, \cdots, v_n\}$ to the (standard) basis S is its coordinate vector with respect to S:

$$P_{B \to S} = [[v_1]_S : [v_2]_S : \cdots : [v_n]_S].$$

Similarly, for a basis $B' = \{w_1, w_2, \cdots, w_n\}$, we have

$$P_{B' \to S} = [[w_1]_S : [w_2]_S : \cdots : [w_n]_S].$$

Then for every $x \in V$, we have

$$P_{B \to S}[x]_B = [x]_S = P_{B' \to S}[x]_{B'}.$$

Therefore we have

$$[x]_{B'} = P^{-1}{}_{B' \to S} P_{B \to S}[x]_B.$$

Since x is arbitrary, we have

$$P^{-1}{}_{B' \to S} P_{B \to S} = P_{B \to B'}.$$

In summary we have

Lemma 3.5.8. Let V be an n-dimensional vector space with bases S, B and B'. Then we have

$$P_{B \to B'} = P^{-1}{}_{B' \to S} P_{B \to S}.$$

When the dimension n is large, $P^{-1}{}_{B' \to S} P_{B \to S}$ can be obtained by Gauss-Jordan elimination:

$$[P_{B' \to S} : P_{B \to S}] \xrightarrow{\text{Elementary row operations}} [I : P^{-1}{}_{B' \to S} P_{B \to S}].$$

Example 3.5.9. Let $B = \{(1, 1), (1, 0)\}$, $B' = \{(0, 1), (1, -1)\}$ be two bases of $\mathbb{R}^2$ (why are they bases?).

1) Find the transition matrix $P_{B \to B'}$;

2) If a vector x has coordinate $[x]_B = (-1, 1)$, find $[x]_{B'}$.

Solution: 1) To find $P_{B \to B'}$, we make use of the standard basis S of $\mathbb{R}^2$. Note that

$$P_{B \to S} = \begin{bmatrix} 1 & 1 \\ 1 & 0 \end{bmatrix}, \quad P_{B' \to S} = \begin{bmatrix} 0 & 1 \\ 1 & -1 \end{bmatrix}.$$

Then we have

$$P_{B \to B'} = P^{-1}{}_{B' \to S} P_{B \to S} = \begin{bmatrix} 0 & 1 \\ 1 & -1 \end{bmatrix} \begin{bmatrix} 0 & 1 \\ 1 & -1 \end{bmatrix} = \begin{bmatrix} 1 & -1 \\ -1 & 2 \end{bmatrix}.$$

2)

$$[x]_{B'} = P_{B \to B'} [x]_B = \begin{bmatrix} 1 & -1 \\ -1 & 2 \end{bmatrix} \begin{bmatrix} 1 \\ -1 \end{bmatrix} = \begin{bmatrix} 2 \\ -3 \end{bmatrix}.$$

$\square$

Remark 3.5.10. (optional) For every vector space V, $\{0\} \subset V$ is a subspace. By convention, we accept that the empty set is a basis for $\{0\}$ and is linearly independent. Hence we say $\dim\{0\} = 0$.

Now we consider a nontrivial vector space V, which has at least one nonzero vector, say $v_1 \neq 0$; then $S_1 = \{v_1\}$ is linearly independent. If $V = \text{span}(S_1)$, then we find a basis for V. Otherwise, there exists $v_2 \notin \text{span}(S_1)$. We add v_2 into S_1 to obtain a linearly independent set of vectors $S_2 = S_1 \cup \{v_2\}$. If $V = \text{span}(S_2)$, then we find a basis for V. Otherwise, we continue to add vectors from $V \setminus \text{span}(S_2)$ to obtain a new linearly independent set of vectors S_3. If the process stops at a finite step n with $V \setminus \text{span}(S_n) = \emptyset$, we obtain a basis with finitely many vectors for the vector space V. Otherwise, the space is infinite dimensional. Certainly this is not an efficient way of finding bases for infinite dimensional spaces and it is often not trivial to find a basis for a specific infinite dimensional space.

Define

$$Y = \text{The collection of all linearly independent subsets of } V,$$

then Y is *partially ordered* by set containment $\subset$, where a partial order is a relation on Y which is

1) reflective: for every $x \in Y$ with $x \subset x$;

2) transitive: for every x, y, $z \in Y$ with $x \subset y$, $y \subset z$ we have $x \subset z$;

3) anti-symmetric: for every x, $y \in Y$ with $x \subset y$, $y \subset x$ we have $x = y$.

If V is an infinite dimensional vector space, then the process of constructing a basis gives a sequence of linearly independent sets:

$$S_1 \subset S_2 \subset \cdots \subset S_n \subset \cdots .$$

Then $X = \{S_n : n \in \mathbb{N}\}$ is a *totally ordered* subcollection of Y. (Every pair in X can be ordered by set containment $\subset$.)

Zorn's lemma in set theory claims that if every totally ordered subset X of a partially ordered set Y has a upper bound, then Y has a maximal element.

If we take union to obtain the upper bound for every totally ordered subcollection such as X, Zorn's lemma applies to our current situation and asserts the existence of a maximal element in Y, which is the maximal linearly independent set of vectors, namely, a basis of V. We arrive at

Theorem 3.5.11. Every vector space has a basis.

The notion of basis we discussed so far is called algebraic basis or **Hamel basis**. Other types of basis of a vector space may be defined when the vector space has extra structures.

Exercise 3.5.12.

1. For the following matrices, determine whether or not i) the rows are linearly dependent; ii) the columns are linearly dependent.

$$a) \ \begin{bmatrix} 1 & 2 & 3 \\ 2 & 3 & 4 \end{bmatrix}, \quad b) \ \begin{bmatrix} 1 & 2 & 3 \\ 2 & 3 & 4 \\ 3 & 4 & 5 \end{bmatrix}, \quad c) \ \begin{bmatrix} 1 & 2 \\ 2 & 3 \\ 3 & 4 \end{bmatrix}.$$

2. Let V be a vector space. Show that if $\{v_1, v_2, v_3, v_4\} \subset V$ is linearly independent, then $\{v_1 - v_2, v_2 - v_3, v_3 - v_4, v_4 - v_1\}$ is linearly independent.

3. Let P_2 be vector space of all polynomials with degree less than or equal to 2. Let $W = \{p \in P_2 : p(1) = 0\}$. Show that W is a subspace of P_2 and find a basis of W.

4. Let P_3 be vector space of all polynomials with degree less than or equal to 3. Let $W = \{p \in P_3 : p(1) = p(2) = 0\}$. Show that W is a subspace of P_3 and find a basis of W.

5. Let $S = \{v_1, v_2, \cdots, v_n\}$ be a set of nonzero vectors in a vector space V with the orthogonal property that

$$v_i \cdot v_j = 0, \ \text{ if } i \neq j.$$

Show that S is linearly independent in V.

6. Let S be the plane $x - 2y + 3z = 0$ in $\mathbb{R}^3$. i) Find the normal of S; ii) Show that S is a subspace of $\mathbb{R}^3$; iii) Show that $\mathbb{R}^3 = S + \text{span}(\mathbf{n})$ and $S \cap \text{span}(\mathbf{n}) = \{\mathbf{0}\}$, where $\mathbf{n}$ is the normal of S. See Exercise 3.1.13 for the definition of set addition.

7. Let S be the plane $x - 2y = 0$ in $\mathbb{R}^3$. i) Find the normal of S; ii) Show that S is a subspace of $\mathbb{R}^3$; iii) Show that $\mathbb{R}^3 = S + \text{span}(\mathbf{n})$ and $S \cap \text{span}(\mathbf{n}) = \{\mathbf{0}\}$, where $\mathbf{n}$ is the normal of S.

8. Let

$$B = \{(1, 1, 1), (1, 1, 0), (1, 0, 0)\}$$

and

$$B' = \{(0,\ 1,\ -1), (1,\ -1,\ 0), (-1,\ 0,\ 0)\}$$

be two bases of $\mathbb{R}^3$ (why are they bases?) i) Find the transition matrix $P_{B \to B'}$; ii) If a vector x has coordinate $[x]_B = (-1,\ 1,\ 0)$, find $[x]_{B'}$.

9. Let

$$B = \{(1,\ 0,\ 1), (0,\ 1,\ 1), (1,\ 1,\ 0)\}$$

and

$$B' = \{(1,\ 0,\ -1), (1,\ -1,\ 0), (-1,\ 0,\ 2)\}$$

be two bases of $\mathbb{R}^3$. i) Find the transition matrix $P_{B \to B'}$; ii) If a vector x has coordinate $[x]_B = (-1,\ 1,\ 1)$, find $[x]_{B'}$.

10. Let

$$P = \begin{bmatrix} 1 & 2 & 3 \\ 2 & 0 & 4 \\ 0 & 4 & 0 \end{bmatrix}, Q = \begin{bmatrix} 1 & 0 & 0 \\ 2 & 1 & 4 \\ 0 & 1 & 0 \end{bmatrix}.$$

Suppose P is the transition matrix from the basis B to the standard basis $S = \{e_1,\ e_2,\ e_3\}$ of $\mathbb{R}^3$, and Q is the transition matrices from the basis B' to the standard basis S. i) Find the transition matrix from B to B'; ii) Find the bases B and B'.

11. Let $f(x) = 2x_1^2 + 2x_2^2 + 2x_3^2 + 4x_1x_2 + 4x_2x_3 + 4x_3x_1$. Find a change of variables $x = Py$, with $x = (x_1,\ x_2,\ x_3)$ and $y = (y_1,\ y_2,\ y_3)$ where P is an invertible 3×3 matrix such that

$$f(Py) = \lambda_1 y_1^2 + \lambda_2 y_2^2 + \lambda_3 y_3^2,$$

for some $\lambda_1,\ \lambda_2,\ \lambda_3 \in \mathbb{R}$.

12. Let A be an $m \times n$ matrix with $m < n$. Show that the columns of A are linearly dependent.

13. Let S be a linearly independent set in a vector space V. If $x \in V$ but $x \notin \operatorname{span}(S)$, then $S \cup \{x\}$ is linearly independent.

14. Let S be a set of vectors in a vector space V. If $x \in S$ and $x \in \operatorname{span}(S \setminus \{x\})$, then $\operatorname{span}(S) = \operatorname{span}(S \setminus \{x\})$.

15. Show that the vector space F of all continuous functions $f : \mathbb{R} \to \mathbb{R}$ is infinite dimensional.

16. Show that every subspace of $\mathbb{R}^n$ is the nullspace of a matrix.

17. Let $\{v_1, v_2, \cdots, v_n\}$ be a basis of $\mathbb{R}^n$. Show that for every $r < n$, $r \in \mathbb{N}$, if $s \neq t$ with $s, t \in \mathbb{R}$, then

$$\text{span}\{v_1, v_2, \cdots v_{r-1}, v_r + sv_n\} \neq \text{span}\{v_1, v_2, \cdots v_{r-1}, v_r + tv_n\}.$$

18. Show that for every $r < n$, $r \in \mathbb{N}$, $\mathbb{R}^n$ has infinitely many r-dimensional subspaces.

Chapter 4

Orthogonality

4.1 Orthogonality of the four subspaces

Example 4.1.1. (Example 3.5.3 revisited.) Let $E = \{x = (x_1, x_2, x_3) \in \mathbb{R}^3 : x_1 - 2x_2 + 3x_3 = 0\}$ be a subset of $\mathbb{R}^3$. Then E is a plane in $\mathbb{R}^3$ passing through $(0, 0, 0)$. For every $x = (x_1, x_2, x_3) \in E$ we have $x_1 = 2x_2 - 3x_3$ and

$$\begin{bmatrix} x_1 \\ x_2 \\ x_3 \end{bmatrix} = \begin{bmatrix} 2x_2 - 3x_3 \\ x_2 \\ x_3 \end{bmatrix} = \begin{bmatrix} 2x_2 \\ x_2 \\ 0 \end{bmatrix} + \begin{bmatrix} 3x_3 \\ 0 \\ x_3 \end{bmatrix} = x_2 \begin{bmatrix} 2 \\ 1 \\ 0 \end{bmatrix} + x_3 \begin{bmatrix} 3 \\ 0 \\ 1 \end{bmatrix}.$$

That is, $E = \text{span}\left(\begin{bmatrix} 2 \\ 1 \\ 0 \end{bmatrix}, \begin{bmatrix} 3 \\ 0 \\ 1 \end{bmatrix} \right)$. Moreover, E can be rewritten as

$$E = \left\{ x = (x_1, x_2, x_3) \in \mathbb{R}^3 : x \perp (1, -2, 3) \right\},$$

which means E is the set of all vectors which is orthogonal to the given vector $(1, -2, 3)$ and hence orthogonal to every vector from the subspace $F = \{k(1, -2, 3) : k \in \mathbb{R}\}$. Namely the subspaces E and F of $\mathbb{R}^3$ are orthogonal. We write $E \perp F$. □

Definition 4.1.2. Let V and W be subspaces of $\mathbb{R}^n$. If for every $(x, y) \in V \times W$ we have
$$x \cdot y = 0,$$
V and W are said to be orthogonal, denoted $V \perp W$.

Remark 4.1.3. For a general real vector space V, orthogonality can be established if we can define a product $\langle \cdot, \cdot \rangle$ between elements which satisfies the following properties,

for every u, v, $w \in V$ and for every scalar $k \in \mathbb{R}$,

1. $\langle u, v \rangle = \langle v, u \rangle$;

2. $\langle u + v, w \rangle = \langle u, v \rangle + \langle v, w \rangle$;

3. $\langle ku, v \rangle = k \langle v, u \rangle$;

4. $\langle v, v \rangle \geq 0$ and $\langle v, v \rangle = 0$ if and only of $v = 0$.

We call $\langle \cdot, \cdot \rangle$ an **inner product** and call V an inner product space. x, $y \in V$ are said to be **orthogonal** if $\langle x, y \rangle = 0$.

For example, for p, $q \in P_n$, with $p(x) = a_0 + a_1 x + a_2 x^2 + \cdots + a_n x^n$ and $q(x) = b_0 + b_1 x + b_2 x^2 + \cdots + b_n x^n$, we can define inner product between p and q by
$$\langle p, q \rangle = a_0 b_0 + a_1 b_1 + \cdots + a_n b_n.$$
Then P_n becomes an inner product space. p, $q \in P_n$ is said to be orthogonal if
$$\langle p, q \rangle = a_0 b_0 + a_1 b_1 + \cdots + a_n b_n = 0.$$
In what follows in this chapter, we assume that a vector space is equipped with an inner product. □

Let A be an $m \times n$ matrix. Consider $N(A) = \{x \in \mathbb{R}^n : Ax = 0\}$. For every $x \in N(A)$ we have
$$\begin{bmatrix} (\text{Row 1 of } A) \cdot x \\ (\text{Row 2 of } A) \cdot x \\ \vdots \\ (\text{Row m of } A) \cdot x \end{bmatrix} = \mathbf{0}.$$
That is, x is orthogonal to every row of A. Therefore, we have
$$N(A) \perp R(A).$$
Recall that the counting theorem says
$$\dim R(A) + \dim N(A) = n.$$

Notice that both $R(A)$ and $N(A)$ are subspaces of $\mathbb{R}^n$. A natural question is that for every $x \in \mathbb{R}^n$ can we split x into two pieces $x_r \in R(A)$ and $x_n \in N(A)$, such that $x = x_r + x_n$? That is, for every $x \in \mathbb{R}^n$, if $x \perp N(A)$, then $x \in R(A)$.

Let us take a basis $S_r = \{v_1, v_2, \cdots, v_r\}$ for $R(A)$, and $S_n = \{w_1, w_2, \cdots, v_{n-r}\}$ for $N(A)$, where $r = \text{rank}(A)$. If $S_r \cup S_n$ is a basis for $\mathbb{R}^n$, then the answer is affirmative. Consider the vector equation

$$c_1 v_1 + c_2 v_2 + \cdots + c_r v_r + d_1 w_1 + d_2 w_2 + \cdots d_{n-r} w_{n-r} = 0. \qquad (4.1)$$

We show that $S_r \cup S_n$ is linearly independent. Otherwise, there exist $(c_1, c_2, \cdots, c_r) \neq 0$ and $(d_1, d_2, \cdots, d_{n-r}) \neq 0$ such that (4.1) holds. Therefore, there exist $x \in R(A)$ and $y \in N(A)$ such that

$$x + y = 0.$$

That is, $x, y \in R(A) \cap N(A)$. Note that we have $R(A) \cap N(A) = \{0\}$, since the only vector orthogonal to itself is the zero vector. Therefore, $x = y = 0$ and

$$c_1 v_1 + c_2 v_2 + \cdots + c_r v_r = 0$$

with $(c_1, c_2, \cdots, c_r) \neq 0$. This is a contradiction since $S_r = \{v_1, v_2, \cdots, v_r\}$ is a basis for $R(A)$. $S_r \cup S_n$ is linearly independent and hence a basis for $\mathbb{R}^n$.

Returning to our original question, for every $x \in \mathbb{R}^n$, it can be represented by the basis $S_r \cup S_n$ as

$$x = c_1 v_1 + c_2 v_2 + \cdots + c_r v_r + d_1 w_1 + d_2 w_2 + \cdots d_{n-r} w_{n-r}. \qquad (4.2)$$

Put $x_r = c_1 v_1 + c_2 v_2 + \cdots + c_r v_r$, $x_n = d_1 w_1 + d_2 w_2 + \cdots d_{n-r} w_{n-r}$. We obtain the split:

$$x = x_r + x_n$$

where $x_r \in R(A)$ and $x_n \in N(A)$. The next question is, is the split unique? Suppose not. We have $x = x_r + x_n = x_r' + x_n'$ where $x_r' \in R(A)$ and $x_n' \in N(A)$. Then we have

$$x_r - x_r' = x_n' - x_n \in R(A) \cap N(A),$$

and $x_r - x_r' = x_n' - x_n = 0$. The split is unique. In summary we have

---

Theorem 4.1.4. Let A be an $m \times n$ matrix. Then for every $x \in \mathbb{R}^n$, there exists a unique orthogonal decomposition

$$x = x_r + x_n,$$

where $x_r \in R(A)$ and $x_n \in N(A)$. That is,

$$\mathbb{R}^n = R(A) \oplus N(A), \quad \text{and} \quad N(A) \perp R(A),$$

where $R(A) \oplus N(A)$ is the **direct sum** of the subspaces $R(A)$ and $N(A)$ of $\mathbb{R}^n$ whose only common element is the zero vector.

Definition 4.1.5. Let W be a subspace of the vector space V. Then $W^\perp$ defined by

$$W^\perp = \{x \in V : x \perp y, \text{ for every } y \in W\}$$

is also a subspace of W and is called the **orthogonal complement** of W.

From Theorem 4.1.4, we know that

$$N(A) = R(A)^\perp, \quad N(A)^\perp = R(A).$$

Example 4.1.6. Let $A = \begin{bmatrix} 1 & 2 \\ 3 & 6 \end{bmatrix}$. We split $x = (4, 3)$ into two parts with $x = x_r + x_n$ where $x_r \in R(A)$ and $x_n \in N(A)$.

Solving $Ax = 0$ we obtain $N(A) = \text{span}\left(\begin{bmatrix} -2 \\ 1 \end{bmatrix}\right)$. Let $x_n = t(-2, 1)$ with $t \neq 0$. Note that

$$x - x_n \perp x_n.$$

We have

$$(x - t(-2, 1)) \cdot t(-2, 1) = 0,$$

which leads to $t = -1$. We have

$$x_n = (2, -1), \; x_r = x - x_n = (2, 4).$$

$\square$

Corollary 4.1.7. Let A be an $m \times n$ matrix. Then for every $x \in \mathbb{R}^m$, there exists a unique orthogonal decomposition

$$x = x_r + x_n,$$

where $x_r \in C(A)$ and $x_n \in N(A^T)$. That is,

$$\mathbb{R}^m = C(A) \oplus N(A^T), \quad \text{and} \quad N(A^T) \perp C(A).$$

An important interpretation of Corollary 4.1.7 is the so-called **Fredholm alternative:**

Corollary 4.1.8. If $b \in \mathbb{R}^m$ is not in the column space of the $m \times n$ matrix A, then it is not orthogonal to $N(A^T)$. Using matrix language, we have
either system

$$Ax = b$$

or system

$$A^T y = 0, \text{ with } y^T b = 1$$

has a solution.

Example 4.1.9. Consider

$$\begin{cases} x + 2y + 3z = 0 \\ 4x + 5y + 6z = 1 \\ 7x + 8y + 9z = 1. \end{cases} \tag{4.3}$$

Let $A = \begin{bmatrix} 1 & 2 & 3 \\ 4 & 5 & 6 \\ 7 & 8 & 9 \end{bmatrix}$ and $b = \begin{bmatrix} 0 \\ 1 \\ 1 \end{bmatrix}$. We have the augmented matrix,

$$[A : b] = \begin{bmatrix} 1 & 2 & 3 & 0 \\ 4 & 5 & 6 & 1 \\ 7 & 8 & 9 & 1 \end{bmatrix} \xrightarrow[R_3 - 7R_1]{R_2 - 4R_1} \begin{bmatrix} 1 & 2 & 3 & 0 \\ 0 & -3 & -6 & 1 \\ 0 & -6 & -12 & 1 \end{bmatrix}$$

$$\xrightarrow{R_3 - 2R_2} \begin{bmatrix} 1 & 2 & 3 & 0 \\ 0 & -3 & -6 & 1 \\ 0 & 0 & 0 & -1 \end{bmatrix},$$

which show that $\operatorname{rank}(A) \neq \operatorname{rank}([A : b])$ and hence $Ax = b$ has no solution. By Corollary 4.1.8, system

$$A^T y = 0, \text{ with } y^T b = 1$$

has a solution. Indeed, we have

$$A^T = \begin{bmatrix} 1 & 4 & 7 \\ 2 & 5 & 8 \\ 3 & 6 & 9 \end{bmatrix} \xrightarrow[R_3 - 3R_1]{R_2 - 2R_1} \begin{bmatrix} 1 & 4 & 7 \\ 0 & -3 & -6 \\ 0 & -6 & -12 \end{bmatrix}$$

$$\xrightarrow{R_3 - 2R_2} \begin{bmatrix} 1 & 4 & 7 \\ 0 & -3 & -6 \\ 0 & 0 & 0 \end{bmatrix}$$

$$\xrightarrow{R_1 + \frac{4}{3}R_2} \begin{bmatrix} 1 & 0 & -1 \\ 0 & -3 & -6 \\ 0 & 0 & 0 \end{bmatrix},$$

which has a one-dimensional nullspace $N(A^T) = \text{span}\left(\begin{bmatrix} 1 \\ -2 \\ 1 \end{bmatrix}\right)$. Let $y =$

$t\begin{bmatrix} 1 \\ -2 \\ 1 \end{bmatrix}$, $t \in \mathbb{R}$ and solve

$$y^T b = 1 \Leftrightarrow t \begin{bmatrix} 1 & -2 & 1 \end{bmatrix} \begin{bmatrix} 0 \\ 1 \\ 1 \end{bmatrix} = 1.$$

We have $t = 1$. That is, $A^T y = 0$ with $y^T b = 1$ has a solution $y = \begin{bmatrix} 1 \\ -2 \\ 1 \end{bmatrix}$.

Exercise 4.1.10.

1. Let $A = \begin{bmatrix} 1 & 2 & 3 \\ 4 & 5 & 6 \\ 7 & 8 & 9 \end{bmatrix}$. i) Find $N(A)$ and split $x = (1,\, 1,\, 1)$ into $x = x_r + x_n$
with $x_n \in N(A)$ and $x_r \in R(A)$; ii) Find the orthogonal complement of $N(A)$.

2. Let $A = \begin{bmatrix} 1 & 2 & 3 \\ 2 & 3 & 4 \\ 3 & 4 & 5 \end{bmatrix}$. i) Find $N(A)$ and split $x = (1,\, 1,\, 1)$ into $x = x_r + x_n$
with $x_n \in N(A)$ and $x_r \in R(A)$; ii) Find the orthogonal complement of $N(A)$.

3. Let S be the plane $x - z = 0$ in $\mathbb{R}^3$. Find the orthogonal complement of S.

4. Let S be the plane $x + y + z = 0$ in $\mathbb{R}^3$. Find the orthogonal complement of S.

5. Show that every triangle inscribed in a semicircle is a right triangle.

6. Show that the function $\langle,\, \rangle : \mathbb{R}^2 \times \mathbb{R}^2 \to \mathbb{R}$ defined by

$$\langle x,\, y \rangle = 2x_1 y_1 - x_2 y_1 - x_1 y_2 + 2x_2 y_2$$

is an inner product on $\mathbb{R}^2$, where $x = (x_1,\, x_2) \in \mathbb{R}^2$ and $y = (y_1,\, x_2) \in \mathbb{R}^2$.

7. Define inner product on the vector space of polynomials P_n by

$$\langle f,\, g \rangle = \int_0^1 f(t)g(t)\mathrm{d}t.$$

i) Show that $W = \{p \in P_4 : p(1) = 0\}$ is a subspace of P_4.

ii) Find the orthogonal complement of W in P_4.

8. Let A and B be subspaces of a real inner product space. Show that

$$A^\perp \cap B^\perp = (A \cup B)^\perp,\ (A \cap B)^\perp = A^\perp \cup B^\perp.$$

9. Let $\beta \in \mathbb{R}^n$ be a nonzero vector. Show that i) $V = \{x : x \cdot \beta = 0\}$ is a subspace of $\mathbb{R}^n$; ii) $\dim V = n - 1$.

10. Let $C([a, b]; \mathbb{R})$ denote the set of all real-valued continuous functions defined on $[a, b]$. i) Show that

$$\langle f, g \rangle = \int_a^b f(t)g(t)\mathrm{d}t$$

defines an inner product on $C([a, b]; \mathbb{R})$.

ii) Show that $\cos mx$ and $\sin nx$ are orthogonal in $C([0, 2\pi]; \mathbb{R})$ if $m \neq n$.

iii) Show that $\cos mx$ and $\cos nx$ are orthogonal in $C([0, 2\pi]; \mathbb{R})$ if $m \neq n$.

iv) Show that $\sin mx$ and $\sin nx$ are orthogonal in $C([0, 2\pi]; \mathbb{R})$ if $m \neq n$.

v) Let $W = \{1, \cos nx, \sin nx\}_{n=1}^N$. Suppose $f \in C([0, 2\pi]; \mathbb{R})$ is a linear combination of the vectors in W. Find the linear combination coefficients.

11. Let $W \subset M_{nn}$ be the set of all skew-symmetric matrices, $V \subset M_{nn}$ be the set of all symmetric matrices. Show that $M_{nn} = V \oplus W$.

12. Let F denote the vector space of all functions $f : \mathbb{R} \to \mathbb{R}$. Let $W \subset F$ be the set of even functions, $V \subset F$ be the set of all even functions. Show that $F = V \oplus W$.

13. Let $A \in M_{nn}$ be such that $A^2 = A$ and $A^T = A$. Show that $\mathbb{R}^n = C(A) \oplus C(I - A)$ and $C(A) \perp C(I - A)$, where $C(A)$ denotes the column space of A.

14. Let $A \in M_{nn}$ be such that $A^2 = -A$ and $A^T = A$. Show that $\mathbb{R}^n = C(A) \oplus C(I + A)$ and $C(A) \perp C(I + A)$, where $C(A)$ denotes the column space of A.

15. Let A be an $n \times n$ matrix with rank n. Suppose A is partitioned into two parts $A = \begin{bmatrix} A_1 \\ A_2 \end{bmatrix}$. Let $W_1 = \{x \in \mathbb{R}^n : A_1 x = 0\}$ and $W_2 = \{x \in \mathbb{R}^n : A_2 x = 0\}$. Show that $\mathbb{R}^n = W_1 \oplus W_2$.

16. Let A be an orthogonal matrix. Define a map $T_A : \mathbb{R}^n \to \mathbb{R}^n$ by $T_A(x) = Ax$. i) Show that for every $x, y \in \mathbb{R}^n$, we have $(Ax) \cdot (Ay) = x \cdot y$; ii) Show that if $V \subset \mathbb{R}^n$ is an **invariant subspace** with respect to T_A, that is, $T_A(V) \subset V$, then $V^\perp$ is also invariant with respect to T_A.

17. Let P_∞ be the set of all polynomials with real coefficients. Define the inner product on P_∞ by

$$\langle f, g \rangle = a_0 b_0 + a_1 b_1 + \cdots + a_r b_r,$$

where $f(x) = a_0 + a_1 x + \cdots + a_m x^m$, $g(x) = b_0 + b_1 x + \cdots + b_n x^n$ and $r = \max\{m, n\}$. Let $V = \{f \in P_\infty : f(0) = 0\}$. i) Show that V is a subspace of P_∞; ii) Find the orthogonal complement of $V^\perp$ in P_∞; iii) Define a map $T : P_\infty \to P_\infty$ by $T(f)(x) = xf(x)$. Show that $T(V) \subset V$ but $T(V^\perp) \not\subset V^\perp$.

18. Verify the Fredholm alternative, for system $Ax = b$ with

$$A = \begin{bmatrix} 1 & 3 & 5 \\ 7 & 9 & 11 \\ 13 & 15 & 17 \end{bmatrix} \text{ and } b = \begin{bmatrix} 0 \\ -2 \\ 1 \end{bmatrix}.$$

4.2 Projections

In the last example of the previous section, we projected an arbitrary vector $x \in \mathbb{R}^n$ onto a line in order to obtain a split $x = x_r + x_n$ such that $x_r \perp x_n$. With such a split we have $\|x\|^2 = \|x_r\|^2 + \|x_n\|^2$ and x_n is the best approximation for x from vectors in the line where x_n lies in. In this section we deepen the understanding of the mechanisms of projections onto general subspaces of $\mathbb{R}^n$ and their matrix representations. In the next section, we discuss how to find the best approximation for a given vector using projections. In what follows, by **projection** of a vector x in a (inner product) vector space V onto a subspace W we mean a map $P : V \to W \subset V$ such that

$$x - Px \perp W.$$

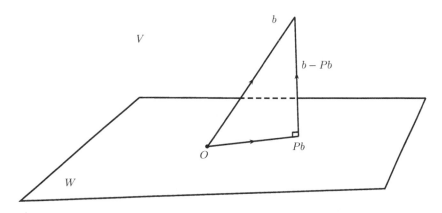

FIGURE 4.1: A projection P from V to the subspace W satisfies $b - Pb \perp W$.

Example 4.2.1. For $b = (b_1, b_2, b_3) \in \mathbb{R}^3$, the projections of x onto subspaces such as the x-axis, y-axis, z-axis, the xy-plane, xz-plane and yz-plane can be obtained by picking out the nonzero coordinates in the subspaces leaving the others zero. Note that to show a vector is orthogonal to a subspace, we need only to show it is orthogonal to every vector of a basis of the subspace.

1) The projection P from $\mathbb{R}^3$ to the x-axis is given by

$$Pb = (b_1, 0, 0),$$

with matrix representation

$$\begin{bmatrix} 1 & 0 & 0 \\ 0 & 0 & 0 \\ 0 & 0 & 0 \end{bmatrix}.$$

Indeed we have $b - Pb = (0, b_2, b_3)$ and the x-axis is spanned by $(1, 0, 0)$. Moreover, $b - Pb \perp (1, 0, 0)$ and $b - Pb$ is orthogonal to the x-axis.

2) The projection P from $\mathbb{R}^3$ to the xy-plane is given by

$$Pb = (0, b_2, b_3),$$

with matrix representation

$$\begin{bmatrix} 1 & 0 & 0 \\ 0 & 1 & 0 \\ 0 & 0 & 0 \end{bmatrix}.$$

Indeed we have $b - Pb = (0, 0, b_3)$ and the xy-plane is spanned by $(1, 0, 0)$ and $(0, 1, 0)$. Moreover $b - Pb \perp (1, 0, 0)$ and $b - Pb \perp (0, 1, 0)$. Hence $b - Pb$ is orthogonal to the xy-plane.

In both cases the projection matrix P satisfies $P^2 = P$.

Projection onto a line

Let $b \in \mathbb{R}^n$ and l be the line spanned by the nonzero vector $a \in \mathbb{R}^n$. Then the projection Pb in l is a scalar multiple of a. Suppose $Pb = \hat{x}a$. With $b - Pb \perp a$ we have $a \cdot (b - \hat{x}a) = 0$ and

$$\hat{x} = \frac{a \cdot b}{a \cdot a} = \frac{a^T b}{a^T a}.$$

That is, $Pb = \hat{x}a = \frac{a^T b}{a^T a}a$. To obtain the representation matrix, we have

$$Pb = \frac{a^T b}{a^T a}a = a\frac{a^T b}{a^T a} = \frac{aa^T b}{a^T a} = \frac{aa^T}{a^T a}b.$$

Lemma 4.2.2. The projection of b onto the line through a in $\mathbb{R}^n$ is

$$Pb = \frac{aa^T}{a^T a}b,$$

and the representation matrix of the projection is

$$P = \frac{aa^T}{a^T a}.$$

Example 4.2.3. Project $b = \begin{bmatrix} 1 \\ 0 \\ -1 \end{bmatrix}$ onto the line l through $a = \begin{bmatrix} 4 \\ 5 \\ 6 \end{bmatrix}$ to find Pb such that $b - Pb \perp l$.

Solution: The projection matrix is

$$P = \frac{aa^T}{a^T a} = \frac{1}{77} \begin{bmatrix} 4 \\ 5 \\ 6 \end{bmatrix} \begin{bmatrix} 4 & 5 & 6 \end{bmatrix} = \frac{1}{77} \begin{bmatrix} 16 & 20 & 24 \\ 20 & 25 & 30 \\ 24 & 30 & 36 \end{bmatrix}.$$

Then $Pb = \frac{1}{77} \begin{bmatrix} 16 & 20 & 24 \\ 20 & 25 & 30 \\ 24 & 30 & 36 \end{bmatrix} \begin{bmatrix} 1 \\ 0 \\ -1 \end{bmatrix} = \frac{1}{77} \begin{bmatrix} -8 \\ -10 \\ -12 \end{bmatrix}$. By Lemma 4.2.2, we have $b - Pb \perp l.$ □

Note the projection matrix $P = \frac{aa^T}{a^T a}$ satisfies

$$P^2 = \frac{aa^T}{a^T a} \frac{aa^T}{a^T a} = \frac{aa^T}{a^T a} = P.$$

Projection onto a subspace

Let $b \in \mathbb{R}^n$ and W be a subspace with a basis $S = \{a_1, a_2, \cdots, a_m\}$. Then the projection Pb in W satisfies that

1) Pb is a linear combination of the basis $S = \{a_1, a_2, \cdots, a_m\}$. That is, there exists $\hat{x} = (\hat{x}_1, \hat{x}_2, \cdots, \hat{x}_m) \in \mathbb{R}^m$ such that

$$Pb = \hat{x}_1 a_1 + \hat{x}_2 a_2 + \cdots + \hat{x}_m a_m = [a_1 : a_2 : \cdots : a_m]\hat{x}.$$

2) $b - Pb$ is orthogonal to W if and only if $b - Pb$ is orthogonal to every vector in the basis $S = \{a_1, a_2, \cdots, a_m\}$. That is,

$$\begin{bmatrix} a_1 \cdot (b - Pb) \\ a_2 \cdot (b - Pb) \\ \vdots \\ a_m \cdot (b - Pb) \end{bmatrix} = \begin{bmatrix} a_1^T \\ a_2^T \\ \vdots \\ a_m^T \end{bmatrix} (b - Pb) = \mathbf{0}.$$

Let $A = [a_1 : a_2 : \cdots : a_m]$. We have $Pb = A\hat{x}$ and $A^T(b - A\hat{x}) = A^T b - A^T A\hat{x} = 0$. Notice that by Theorem 2.5.6, $A^T A$ is invertible since its columns are a basis and hence linearly independent. Therefore, we have

$$\hat{x} = (A^T A)^{-1} A^T b,$$

and

$$Pb = A\hat{x} = A(A^T A)^{-1} A^T b.$$

In summary, we have

> **Lemma 4.2.4.** Let W be a subspace of $\mathbb{R}^n$ with a basis $S = \{a_1, a_2, \cdots, a_m\}$ and $A = [a_1 : a_2 : \cdots : a_m]$. Then for every $b \in \mathbb{R}$, the projection of b onto the subspace W is
>
> $$Pb = A(A^T A)^{-1} A^T b,$$
>
> and the representation matrix of the projection is
>
> $$P = A(A^T A)^{-1} A^T.$$

Example 4.2.5. Let

$$A = \begin{bmatrix} 1 & 1 & 1 \\ 1 & 0 & 0 \\ 0 & 1 & 0 \\ 0 & 0 & 1 \end{bmatrix}, b = \begin{bmatrix} 1 \\ 1 \\ 1 \\ 1 \end{bmatrix}.$$

Find the orthogonal projection Pb of b onto the column space of A.
Solution: We notice that the columns of A are linearly independent because $\text{rank}(A) = 3$ and A has exactly three columns. Then Lemma 4.2.4 applies. We solve the system $A^T A \hat{x} = A^T b$ where

$$A^T A = \begin{bmatrix} 1 & 1 & 0 & 0 \\ 1 & 0 & 1 & 0 \\ 1 & 0 & 0 & 1 \end{bmatrix} \begin{bmatrix} 1 & 1 & 1 \\ 1 & 0 & 0 \\ 0 & 1 & 0 \\ 0 & 0 & 1 \end{bmatrix} = \begin{bmatrix} 2 & 1 & 1 \\ 1 & 2 & 1 \\ 1 & 1 & 2 \end{bmatrix}, A^T b = \begin{bmatrix} 2 \\ 2 \\ 2 \end{bmatrix}.$$

We use Gauss-Jordan elimination to determine the inverse of $A^T A$.

$$[A^T A : I] = \begin{bmatrix} 2 & 1 & 1 & 1 & 0 & 0 \\ 1 & 2 & 1 & 0 & 1 & 0 \\ 1 & 1 & 2 & 0 & 0 & 1 \end{bmatrix}$$

$$\xRightarrow{R_1 + R_2 + R_3} \begin{bmatrix} 4 & 4 & 4 & 1 & 1 & 1 \\ 1 & 4 & 1 & 0 & 1 & 0 \\ 1 & 1 & 4 & 0 & 0 & 1 \end{bmatrix}$$

$$\xRightarrow{R_1/4} \begin{bmatrix} 1 & 1 & 1 & \frac{1}{4} & \frac{1}{4} & \frac{1}{4} \\ 1 & 2 & 1 & 0 & 1 & 0 \\ 1 & 1 & 2 & 0 & 0 & 1 \end{bmatrix}$$

$$\xRightarrow[R_3 - R_1]{R_2 - R_1} \begin{bmatrix} 1 & 1 & 1 & \frac{1}{4} & \frac{1}{4} & \frac{1}{4} \\ 0 & 1 & 0 & \frac{-1}{4} & \frac{3}{4} & \frac{-1}{4} \\ 0 & 0 & 1 & \frac{-1}{4} & \frac{-1}{4} & \frac{3}{4} \end{bmatrix}$$

$$\xRightarrow[R_1 - R_3]{R_1 - R_2} \begin{bmatrix} 1 & 0 & 0 & \frac{3}{4} & \frac{-1}{4} & \frac{-1}{4} \\ 0 & 1 & 0 & \frac{-1}{4} & \frac{3}{4} & \frac{-1}{4} \\ 0 & 0 & 1 & \frac{-1}{4} & \frac{-1}{4} & \frac{3}{4} \end{bmatrix}.$$

Then the inverse of $A^T A$ is

$$(A^T A)^{-1} = \begin{bmatrix} \frac{3}{4} & \frac{-1}{4} & \frac{-1}{4} \\ \frac{-1}{4} & \frac{3}{4} & \frac{-1}{4} \\ \frac{-1}{4} & \frac{-1}{4} & \frac{3}{4} \end{bmatrix}.$$

Therefore, the orthogonal projection Pb of b onto the column spaces of A is

$$Pb = A\hat{x} = A(A^T A)^{-1} A^T b = \begin{bmatrix} 1 & 1 & 1 \\ 1 & 0 & 0 \\ 0 & 1 & 0 \\ 0 & 0 & 1 \end{bmatrix} \begin{bmatrix} \frac{3}{4} & \frac{-1}{4} & \frac{-1}{4} \\ \frac{-1}{4} & \frac{3}{4} & \frac{-1}{4} \\ \frac{-1}{4} & \frac{-1}{4} & \frac{3}{4} \end{bmatrix} \begin{bmatrix} 2 \\ 2 \\ 2 \end{bmatrix} = \begin{bmatrix} \frac{3}{2} \\ \frac{1}{2} \\ \frac{1}{2} \\ \frac{1}{2} \end{bmatrix}.$$

Remark 4.2.6. We remark that the projection matrix $P = A(A^T A)^{-1} A^T$ satisfies

$$P^2 = A(A^T A)^{-1} A^T A(A^T A)^{-1} A^T = A(A^T A)^{-1} A^T = P,$$
$$P^T = (A(A^T A)^{-1} A^T)^T = A(A^T A)^{-1} A^T = P.$$

That is,

> $P^2 = P$ and P is symmetric.

Then we have

$$P(b - Pb) = Pb - P^2 b = Pb - Pb = 0,$$

which imply that $b - Pb \in N(P)$. The symmetry of P implies that the row space of P is the same as the column space of P. Then for every $b \in \mathbb{R}^n$, b has an orthogonal decomposition

$$b = b_n + b_r,$$

with $b_n = b - Pb \in N(P)$ and $b_r = Pb \in C(P) = R(P)$.

Exercise 4.2.7.

1. Let $W = \text{span}\left(\begin{bmatrix} 1 \\ -1 \\ 2 \end{bmatrix}, \begin{bmatrix} 0 \\ 1 \\ 0 \end{bmatrix} \right)$. Find the projection matrix P for the projection onto W.

2. Find the orthogonal projection of the vector $b = (1, 2, 4) \in \mathbb{R}^3$ onto the line

$$l : \frac{x}{2} = \frac{y}{3} = \frac{z}{4},$$

and find the distance from the point $(1, 2, 4)$ to the line l.

3. Find the orthogonal projection of the vector $b = (1, 2, 4) \in \mathbb{R}^3$ onto the plane $S : x - y + z = 0$ and find the distance from the point $(1, 2, 4)$ to the plane S.

4. Find the orthogonal projection of the vector $b = (1, 0, 4) \in \mathbb{R}^3$ onto the plane $S : x + y = 0$ and find the distance from the point $(1, 0, 4)$ to the plane S.

5. Find the distance from $\alpha = (-1, 1, 1, 0)$ to $W = \text{span}(\beta_1, \beta_2, \beta_3)$ in $\mathbb{R}^4$, where $\beta_1 = (1, 0, 0, 1)$, $\beta_2 = (1, 1, 0, 2)$ and $\beta_3 = (1, 1, 4, 1)$.

6. Find the distance from $\alpha = (-1, 1, 1, 0)$ to $W = \text{span}(\beta_1, \beta_2, \beta_3)$ in $\mathbb{R}^4$, where $\beta_1 = (1, 0, 0, 1)$, $\beta_2 = (1, 1, 0, 2)$ and $\beta_3 = (0, 1, 0, 1)$.

7. Let u and v be vectors in $\mathbb{R}^3$. Show that

$$(u + v) \cdot (u - v) = \|u\|^2 - \|v\|^2.$$

8. Show that the diagonals of a parallelogram in $\mathbb{R}^3$ with equal sides are perpendicular to each other.

9. Let $A(0, 1, 1)$, $B(1, 1, 1)$ and $C(-1, -1, 0)$ be the vertices of the triangle $\triangle ABC$. Find the area of $\triangle ABC$.

10. Let u and v be nonzero vectors in $\mathbb{R}^3$. Show that the area of the triangle determined by u and v is

$$\frac{1}{2} \sqrt{\|u\|^2 \|v\|^2 - (u \cdot v)^2}.$$

11. (Parallelogram law) Let u and v be vectors in $\mathbb{R}^n$. Show that

$$2(\|u\|^2 + \|v\|^2) = \|u + v\|^2 + \|u - v\|^2.$$

12. Let A be an $m \times n$ matrix with linearly independent rows. Find the representation matrix P of the orthogonal projection onto the row space of A.

13. Let s and l be lines in $\mathbb{R}^3$ given by

$$s : \frac{x - 1}{2} = \frac{y - 1}{3} = \frac{z - 1}{4}, \quad l : \frac{x}{2} = \frac{y}{2} = \frac{z}{2}.$$

Find the distance d between s and l:

$$d = \min_{p \in s,\, q \in l} \|p - q\|.$$

14. Let W be a subspace of $\mathbb{R}^n$. Show that the orthogonal projection $P : V \to W$ is a linear map.

15. What is the representation matrix P for the projection from $\mathbb{R}^n$ onto itself?

16. Let P be the representation matrix for the projection onto a (true) subspace W of $\mathbb{R}^n$. i) For every $x \notin W$, is it possible $Px = x$? ii) Is P invertible? Explain your answer.

17. Let P be the representation matrix for the projection onto a (true) subspace W of $\mathbb{R}^n$. Is P dependent on your choice of the basis for W? Explain your answer.

18. Let P be an $n \times n$ matrix and satisfy $P^T = P$ and $P^2 = P$. Is it true that P is the representation matrix for the projection onto a (true) subspace W of $\mathbb{R}^n$? Namely, for every $b \in \mathbb{R}^n$, is $b - Pb$ orthogonal to Pb?

19. Let P be the representation matrix for the projection onto a subspace W of $\mathbb{R}^n$. Is it true that W equals the column space of P?

20. Let $C([0, 2\pi]; \mathbb{R})$ denote the set of all real-valued continuous functions defined on $[0, 2\pi]$. Let $W = \text{span}\{1, \cos nx, \sin nx\}_{n=1}^N$. Define an inner product on $C([0, 2\pi]; \mathbb{R})$ by

$$\langle f, g \rangle = \int_0^{2\pi} f(t)g(t)\mathrm{d}t, \quad f, g \in C([0, 2\pi]; \mathbb{R}).$$

i) For every $f \in C([0, 2\pi]; \mathbb{R})$, find the orthogonal projection f_p of f onto W;

ii) For every $f \in C([0, 2\pi]; \mathbb{R})$, find the distance d of f to W defined by

$$d = \langle f - f_p, f - f_p \rangle^{\frac{1}{2}},$$

where f_p is the orthogonal projection of f onto W.

iii) Let $f_0(x) = 1 + x^2$ be a function in $C([0, 2\pi]; \mathbb{R})$. Find the orthogonal projection f_{p_0} of f_0 onto W and $\langle f_0 - f_{p_0}, f_0 - f_{p_0} \rangle^{\frac{1}{2}}$.

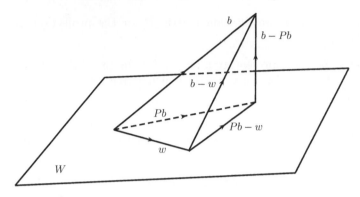

FIGURE 4.2: $\|b - Pb\| \leq \|b - w\|$ for every $w \in W$.

4.3   Least squares approximations

We have learned that a linear system $Ax = b$ does not have a solution if $b \notin C(A)$. Let $W = C(A)$. A natural question is how to find a $b_r \in W$ which best approximates b? Geometrically, the answer is the projection $b_r = Pb$ of b onto W, where P is the representation matrix of the projection onto W. Indeed, for every $w \in W$,

$$b - w = b - Pb + Pb - w,$$

where $(b - Pb) \perp (Pb - w)$ since $Pb - w \in W$ and $(b - Pb) \perp W$. By the Pythagorean theorem, we have

$$\|b - w\|^2 = \|b - Pb\|^2 + \|Pb - w\|^2,$$

which implies that

$$\|b - Pb\| \leq \|b - w\|, \text{ for every } w \in W,$$

and the equality happens only if $w = Pb$.

Theorem 4.3.1. Let W be a subspace of $\mathbb{R}^n$ and P the representation matrix of the projection onto W. Then for every $b \in \mathbb{R}$,

$$\|b - Pb\| \leq \|b - w\|, \text{ for every } w \in W,$$

and the equality happens only if $w = Pb$.

Definition 4.3.2. Let A be an $n \times m$ matrix and $b \in \mathbb{R}^n$. We call $\hat{x} \in \mathbb{R}^m$ a **least squares solution** to system $Ax = b$ if $A\hat{x} = Pb$, where Pb is the projection of b onto the column space of A.

Certainly $A\hat{x} = Pb$ is always consistent with at least one solution since Pb is by definition in the column space of A. The question is how to find $\hat{x}$ without having to find the representation matrix P?

Let us first find a necessary condition that $\hat{x}$ must satisfy. Notice that by definition of projection we have $b - Pb \perp C(A) = R(A^T)$. That is, $b - Pb$ is orthogonal to each row of A^T. Therefore we have

$$A^T(b - Pb) = A^T(b - A\hat{x}) = 0,$$

which is equivalent to

$$A^T A\hat{x} = A^T b. \tag{4.4}$$

That is, if $\hat{x}$ is a least squares solution, it must satisfy (4.4) which we call the **normal system** corresponding to $Ax = b$. This also shows that the normal system $A^T A\hat{x} = A^T b$ is always consistent.

A question remaining is that if $\hat{x}$ satisfies the normal system (4.4), is it a solution to $A\hat{x} = Pb$ and hence a least squares solution? Indeed, if $\hat{x}$ satisfies the normal system (4.4), $b - A\hat{x}$ is orthogonal to $C(A) = R(A^T)$. Then we have an orthogonal decomposition of b:

$$b = b - A\hat{x} + A\hat{x}.$$

Another orthogonal decomposition of b is

$$b = b - Pb + Pb.$$

By uniqueness of orthogonal decomposition with respect to the subspace $W = C(A)$, we have

$$A\hat{x} = Pb,$$

which means that every solution of the normal system is a least squares solution.

Theorem 4.3.3. Let A be an $n \times m$ matrix and $b \in \mathbb{R}^n$. $\hat{x} \in \mathbb{R}^m$ is a least squares solution to system $Ax = b$ if and only if $A^T A\hat{x} = A^T b$. Moreover, $A^T A\hat{x} = A^T b$ is always consistent.

Corollary 4.3.4. Let A be an $n \times m$ matrix and $b \in \mathbb{R}^n$. If the columns of A are linearly independent, then the least squares solution to system $Ax = b$ is

$$\hat{x} = (A^T A)^{-1} A^T b.$$

Example 4.3.5. (Example 4.2.5 revisited.) Let

$$A = \begin{bmatrix} 1 & 1 & 1 \\ 1 & 0 & 0 \\ 0 & 1 & 0 \\ 0 & 0 & 1 \end{bmatrix}, \quad b = \begin{bmatrix} 1 \\ 1 \\ 1 \\ 1 \end{bmatrix}.$$

Find a least squares solution to $Ax = b$.

Solution: We solve the system $A^T A \hat{x} = A^T b$ where

$$A^T A = \begin{bmatrix} 1 & 1 & 0 & 0 \\ 1 & 0 & 1 & 0 \\ 1 & 0 & 0 & 1 \end{bmatrix} \begin{bmatrix} 1 & 1 & 1 \\ 1 & 0 & 0 \\ 0 & 1 & 0 \\ 0 & 0 & 1 \end{bmatrix} = \begin{bmatrix} 2 & 1 & 1 \\ 1 & 2 & 1 \\ 1 & 1 & 2 \end{bmatrix}, \quad A^T b = \begin{bmatrix} 2 \\ 2 \\ 2 \end{bmatrix}.$$

We use Gauss–Jordan elimination to determine the inverse of $A^T A$ and obtain

$$(A^T A)^{-1} = \begin{bmatrix} \frac{3}{4} & \frac{-1}{4} & \frac{-1}{4} \\ \frac{-1}{4} & \frac{3}{4} & \frac{-1}{4} \\ \frac{-1}{4} & \frac{-1}{4} & \frac{3}{4} \end{bmatrix}.$$

Therefore, the least squares solution $\hat{x}$ of $Ax = b$ is

$$\hat{x} = (A^T A)^{-1} A^T b = \begin{bmatrix} \frac{3}{4} & \frac{-1}{4} & \frac{-1}{4} \\ \frac{-1}{4} & \frac{3}{4} & \frac{-1}{4} \\ \frac{-1}{4} & \frac{-1}{4} & \frac{3}{4} \end{bmatrix} \begin{bmatrix} 2 \\ 2 \\ 2 \end{bmatrix} = \begin{bmatrix} \frac{1}{2} \\ \frac{1}{2} \\ \frac{1}{2} \end{bmatrix}.$$

Example 4.3.6. Let (x_1, y_1), (x_2, y_2), $\cdots$, (x_m, y_m) be a set of data for the pair $(x, y) \in \mathbb{R}^2$. Suppose that x and y are related by a polynomial

$$y = p(x) = a_0 + a_1 x + a_2 x^2 \cdots + a_n x^n.$$

We wish to use the observed data to solve for the unknown coefficients $a_0, a_1, \cdots, a_n$ and set up the following equations

$$\begin{cases} y_1 = p(x_1) \\ y_2 = p(x_2) \\ \vdots \\ y_n = p(x_n), \end{cases} \Leftrightarrow \begin{cases} y_1 = a_0 + a_1 x_1 + a_2 x_2^2 \cdots + a_n x_1^n \\ y_2 = a_0 + a_1 x_2 + a_2 x_2^2 \cdots + a_n x_2^n \\ \vdots \\ y_m = a_0 + a_1 x_m + a_2 x_m^2 \cdots + a_n x_m^n \end{cases},$$

which can be written as $Au = b$, where

$$A = \begin{bmatrix} 1 & x_1 & x_1^2 & \cdots & x_1^n \\ 1 & x_2 & x_2^2 & \cdots & x_2^n \\ \vdots & \vdots & \vdots & & \vdots \\ 1 & x_m & x_m^2 & \cdots & x_m^n \end{bmatrix}, \quad u = \begin{bmatrix} a_0 \\ a_1 \\ \vdots \\ a_n \end{bmatrix}, \quad b = \begin{bmatrix} y_1 \\ y_2 \\ \vdots \\ y_n \end{bmatrix}.$$

Such an inhomogeneous system may not have a solution for u, especially when the system is over-determined with $m > n$. However, we can find a least squares solution $\hat{u}$ of $Au = b$ which minimizes $\|b - Au\|$. Namely, a least squares solution $\hat{u}$ of $Au = b$ is a solution to the minimization problem

$$\min_{u \in \mathbb{R}^n} \|b - Au\|,$$

and

$$\|b - A\hat{u}\| = \min_{u \in \mathbb{R}^n} \|b - Au\|.$$

By Theorem 4.3.3, every solution of $A^T A\hat{u} = A^T b$ is a least squares solution to $Au = b$. We remark that the process of using a set of data to fit in a polynomial is called **polynomial interpolation.**

Exercise 4.3.7.

1. Is it true a least squares solution may not be unique? Explain your answer.

2. Let A be an $m \times n$ matrix and b is $n \times 1$. Is it true a least squares solution to $Ax = b$ always exists? Explain your answer.

3. Let x_0 be a least squares solution of $Ax = b$. Show that $b^T Ax_0 \geq 0$.

4. Let

$$A = \begin{bmatrix} 1 & x_1 & x_1^2 & \cdots & x_1^n \\ 1 & x_2 & x_2^2 & \cdots & x_2^n \\ \vdots & & \vdots & & \vdots \\ 1 & x_m & x_m^2 & \cdots & x_m^n \end{bmatrix}.$$

Suppose that there are at least $n+1$ distinct values among $\{x_1, x_2, \cdots, x_m\}$. Show that the columns of A are linearly independent. (Hint: check the maximal number of distinct roots of a nontrivial polynomial $p(x) = a_0 + a_1 x + a_2 x^2 + \cdots + a_n x^n$. A is called a **Vandermonde matrix.**)

5. Use the polynomial $h = bt + c$ to fit in the data in the sense of least squares:

h	3	2	1	2	0
t	0	1	2	3	4

6. Use the polynomial $h = at^2 + bt + c$ to fit in the data in the sense of least squares:

h	3	2	1	2	0
t	0	1	2	3	4

7. Let $Ax = b$ be consistent. i) Is it true that every least squares solution of $Ax = b$ is also a solution of $Ax = b$? Justify your answer. ii) Is it true that every solution of $Ax = b$ is also a least squares solution of $Ax = b$? Justify your answer.

8. Let $\mathbf{n} = (1, -1, 1, 0)$ be the normal of the hyperplane (see Exercise 3.1.13 for the definition) S in $\mathbb{R}^4$ passing through the point $(1, 2, 3, 4)$. i) Find the equation for S; ii) Find the distance from $b = (1, 2, 4, 5)$ to S.

9. Show that if u_0 is a solution of

$$\min_{u \in \mathbb{R}^n} \|b - Au\|,$$

then $A^T A u_0 = A^T b$. That is, u_0 is a least squares solution of $Ax = b$.

10. Show that if u_0 is a least squares solution of $Ax = b$, then u_0 is a solution of

$$\min_{u \in \mathbb{R}^n} \|b - Au\|.$$

4.4 Orthonormal bases and Gram–Schmidt

Orthonormal bases

We know that if W is a subspace of $\mathbb{R}^n$ with basis $S = \{a_1, a_2, \cdots, a_m\}$, then the projection onto W has a representation matrix

$$P = A(A^T A)^{-1} A^T,$$

where $A = [a_1 : a_2 : \cdots : a_m]$. The real application of the matrix P will involve the computation of the inverse of the $m \times m$ matrix $A^T A$, which is usually not convenient. It would be a big relief if $A^T A = I$, namely, if A is an orthogonal matrix (see Definition 2.5.11). Indeed, if A is an orthogonal matrix, we have the following observations:

1) $A^T A = I$ implies that the columns of A are unit vectors and are orthogonal to each other:

$$a_i^T a_j = \begin{cases} 1 & \text{if } i = j \\ 0 & \text{if } i \neq j. \end{cases} \tag{4.5}$$

We call $S = \{a_1, a_2, \cdots, a_m\}$ an **orthonormal basis** if (4.5) is satisfied.

2) The representation matrix of the projection onto W is simplified to $P = AA^T$.

3) For every $b \in \mathbb{R}^n$, its orthogonal projection onto W is

$$Pb = AA^T b = [a_1 : a_2 : \cdots : a_m] \begin{bmatrix} a_1^T \\ a_2^T \\ \vdots \\ a_m^T \end{bmatrix} b = a_1 a_1^T b + a_2 a_2^T b + \cdots + a_m a_m^T b.$$

Note that $a_i a_i^T b = \frac{a_i a_i^T}{a_i^T a_i} b$ is exactly the projection of b onto the one-dimensional subspace of W spanned by a_i, $i = 1, 2, \cdots, m$. This implies that if W has an orthonormal basis, namely, if a basis consists of unit and orthogonal vectors, the projection onto W is the sum of the projections along the individual vectors in the orthonormal basis.

Orthogonal basis

Suppose that the subspace W has an orthogonal basis $S = \{a_1, a_2, \cdots, a_m\}$, which are not necessarily unit vectors. Namely, $A^T A$ is a diagonal matrix with

$$a_i^T a_j = \begin{cases} \neq 0 & \text{if } i = j \\ 0 & \text{if } i \neq j. \end{cases} \tag{4.6}$$

We call $S = \{a_1, a_2, \cdots, a_m\}$ an **orthogonal basis** if (4.6) is satisfied. In this case, the orthogonal projection onto W has an easy representation. In fact,

$$A^T A = \begin{bmatrix} a_1^T a_1 & 0 & \cdots & 0 \\ 0 & a_2^T a_2 & \cdots & 0 \\ \vdots & \vdots & \ddots & \vdots \\ 0 & 0 & \vdots & a_m^T a_m \end{bmatrix}.$$

For every $b \in \mathbb{R}^n$, its orthogonal projection onto W is

$$Pb = A(A^T A)^{-1} A^T b = [a_1 : a_2 : \cdots : a_m] \begin{bmatrix} \frac{1}{a_1^T a_1} & 0 & \cdots & 0 \\ 0 & \frac{1}{a_2^T a_2} & \cdots & 0 \\ \vdots & \vdots & \ddots & \vdots \\ 0 & 0 & \vdots & \frac{1}{a_m^T a_m} \end{bmatrix} \begin{bmatrix} a_1^T \\ a_2^T \\ \vdots \\ a_m^T \end{bmatrix} b$$

$$= \frac{a_1 a_1^T}{a_1^T a_1} b + \frac{a_2 a_2^T}{a_2^T a_2} b + \cdots + \frac{a_m a_m^T}{a_m^T a_m} b$$

$$= \frac{a_1^T b}{a_1^T a_1} a_1 + \frac{a_2^T b}{a_2^T a_2} a_2 + \cdots + \frac{a_m^T b}{a_m^T a_m} a_m.$$

Similar to the case of orthonormal basis, $Pb = \frac{a_1 a_1^T}{a_1^T a_1} b + \frac{a_2 a_2^T}{a_2^T a_2} b + \cdots + \frac{a_m a_m^T}{a_m^T a_m} b$ implies that the projection onto W is the sum of the projections along the individual vectors in the orthogonal basis. $Pb = \frac{a_1^T b}{a_1^T a_1} a_1 + \frac{a_2^T b}{a_2^T a_2} a_2 + \cdots + \frac{a_m^T b}{a_m^T a_m} a_m$ indicates that the coefficient of the decomposition for Pb along the a_i direction is $\frac{a_i^T b}{a_i^T a_i}$. In summary, we have

Lemma 4.4.1. Let W be a subspace of $\mathbb{R}^n$ with an orthogonal basis $S = \{a_1, a_2, \cdots, a_m\}$. For every $b \in \mathbb{R}^n$, its orthogonal projection onto W is

$$Pb = \left(\frac{a_1 a_1^T}{a_1^T a_1} + \frac{a_2 a_2^T}{a_2^T a_2} + \cdots + \frac{a_m a_m^T}{a_m^T a_m} \right) b$$

$$= \frac{a_1^T b}{a_1^T a_1} a_1 + \frac{a_2^T b}{a_2^T a_2} a_2 + \cdots + \frac{a_m^T b}{a_m^T a_m} a_m.$$

If S is orthonormal, then

$$Pb = \left(a_1 a_1^T + a_2 a_2^T + \cdots + a_m a_m^T \right) b$$

$$= (a_1^T b) a_1 + (a_2^T b) a_2 + \cdots + (a_m^T b) a_m.$$

Note that if $W = \mathbb{R}^n$, the projection is just the identity map with $P = I$. For every $b \in \mathbb{R}^n$, its orthogonal projection onto W is

$$b = \frac{a_1^T b}{a_1^T a_1} a_1 + \frac{a_2^T b}{a_2^T a_2} a_2 + \cdots + \frac{a_m^T b}{a_m^T a_m} a_m,$$

which is a orthogonal decomposition of b along each of the vectors in the orthogonal basis $S = \{a_1, a_2, \cdots, a_m\}$.

A natural question next is how to find orthogonal and orthonormal basis from a given basis? The Gram-Schmidt process is a procedure for this purpose. Before we detail the process of creating orthogonal basis, let us first observe that a set of nonzero orthogonal vectors $S = \{a_1, a_2, \cdots, a_m\}$ in $\mathbb{R}^n$ are linearly independent. Indeed, consider the vector equation

$$c_1 a_1 + c_2 a_2 + \cdots c_m a_m = 0.$$

Multiply both sides by a_i^T, $i = 1, 2, \cdots, m$. We have

$$a_i^T (c_1 a_1 + c_2 a_2 + \cdots + c_m a_m) = 0 \Rightarrow c_i a_i^T a_i = 0,$$

which leads to $c_i = 0$ since a_i is nonzero. That is

Lemma 4.4.2. If $S = \{a_1, a_2, \cdots, a_m\}$ is a set of orthogonal nonzero vectors in $\mathbb{R}^n$, then S is linearly independent.

Gram–Schmidt orthogonalization process

Let W be a subspace of $\mathbb{R}^n$ with an orthogonal basis $S = \{a_1, a_2, \cdots, a_m\}$. We want to convert it into a new basis $S' = \{b_1, b_2, \cdots, b_m\}$. The idea is that for every $i = 1, 2, \cdots, m$ and beginning with $b_1 = a_1$, use orthogonal projection to find b_{i+1} such that

$$b_{i+1} \perp \text{span}\{a_1, a_2, \cdots, a_i\},$$

and that
$$\text{span}\{a_1, a_2, \cdots, a_{i+1}\} = \text{span}\{b_1, b_2, \cdots, b_{i+1}\}.$$

When the process terminates, we obtain a set of orthogonal basis $S' = \{b_1, b_2, \cdots, b_m\}$ for W. To be specific, we carry out the following steps.

1) Let $b_1 = a_1$;

2) Find an orthogonal decomposition of a_2 with respect to $\text{span}\{b_1\} = \text{span}\{a_1\}$ and let b_2 be the component of a_2 orthogonal to $\text{span}\{b_1\} = \text{span}\{a_1\}$:

$$b_2 = a_2 - \frac{b_1 b_1^T}{b_1^T b_1} a_2 = a_2 - \frac{b_1^T a_2}{b_1^T b_1} b_1.$$

Then $b_2 \neq 0$ since $a_2 \notin \text{span}\{a_1\} = \text{span}\{b_1\}$. Then $\{b_1, b_2\}$ is linearly independent since $b_2 \perp \text{span}\{b_1\}$. Moreover the orthogonal set $\{b_1, b_2\}$ is contained in the two-dimensional space $\text{span}\{a_1, a_2\}$. Therefore, we have

$$\text{span}\{a_1, a_2\} = \text{span}\{b_1, b_2\}.$$

3) Find an orthogonal decomposition of a_3 with respect to $\text{span}\{b_1, b_2\} = \text{span}\{a_1, a_2\}$ and let b_3 be the component of a_3 orthogonal to $\text{span}\{b_1, b_2\} = \text{span}\{a_1, a_2\}$:

$$b_3 = a_3 - \frac{b_1 b_1^T}{b_1^T b_1} a_3 - \frac{b_2 b_2^T}{b_2^T b_2} a_3 = a_3 - \frac{b_1^T a_3}{b_1^T b_1} b_1 - \frac{b_2^T a_3}{b_2^T b_2} b_2.$$

Then $b_3 \neq 0$ since $a_3 \notin \text{span}\{a_1, a_2\} = \text{span}\{b_1, b_2\}$. Then $\{b_1, b_2, b_3\}$ is linearly independent since $b_3 \perp \text{span}\{b_1, b_2\}$. Moreover the orthogonal set $\{b_1, b_2, b_3\}$ is contained in the three-dimensional space $\text{span}\{a_1, a_2, a_3\}$. Therefore, we have

$$\text{span}\{a_1, a_2, a_3\} = \text{span}\{b_1, b_2, b_3\}.$$

4) Successively find an orthogonal decomposition of a_{m+1} with respect to

$$\text{span}\{b_1, b_2, \cdots, b_m\} = \text{span}\{a_1, a_2, \cdots, a_m\},$$

and let b_{m+1} be the component of a_{m+1} which is orthogonal to $\text{span}\{b_1, b_2, \cdots, b_m\}$:

$$b_{m+1} = a_{m+1} - \frac{b_1 b_1^T}{b_1^T b_1} a_{m+1} - \frac{b_2 b_2^T}{b_2^T b_2} a_{m+1} - \cdots - \frac{b_m b_m^T}{b_m^T b_m} a_{m+1} \tag{4.7}$$

$$= a_{m+1} - \frac{b_1^T a_{m+1}}{b_1^T b_1} b_1 - \frac{b_2^T a_{m+1}}{b_2^T b_2} b_2 - \cdots - \frac{b_m^T a_{m+1}}{b_m^T b_m} b_m. \tag{4.8}$$

Then we have $b_{m+1} \neq 0$ since

$$a_{m+1} \notin \text{span}\{a_1, a_2, \cdots, a_m\} = \text{span}\{b_1, b_2, \cdots, b_m\}.$$

We obtain that $\{b_1, b_2, \cdots, b_m\}$ is linearly independent since

$$b_{m+1} \perp \text{span}\{b_1, b_2, \cdots, b_m\}.$$

Moreover the orthogonal set $\{b_1, b_2, \cdots, b_{m+1}\}$ is contained in the $m+1$-dimensional space $\text{span}\{b_1, b_2, \cdots, b_{m+1}\}$. Therefore, we have

$$\text{span}\{a_1, a_2, \cdots, a_{m+1}\} = \text{span}\{b_1, b_2, \cdots, b_{m+1}\}.$$

Example 4.4.3. Let $W = \text{span}\{w_1, w_2, w_3\}$ be a subspace of $\mathbb{R}^3$, where

$$w_1 = (1, -1, 0), w_2 = (-2, 3, 1), w_3 = (1, 2, 4).$$

i) Show that $\{w_1, w_2, w_3\}$ is linearly independent. Then use the Gram-Schmidt process to convert $\{w_1, w_2, w_3\}$ into an orthogonal basis $V = \{v_1, v_2, v_3\}$ of W;

ii) Let $u = (1, 1, 2)$. Find the coordinate vector $[u]_V$ relative to the orthogonal basis V obtained in i).

Solution: i) To show that W is linearly independent, we consider the vector equation

$$c_1 w_1 + c_2 w_2 + c_3 w_3 = 0,$$

which is equivalent to

$$\begin{bmatrix} 1 & -2 & 1 \\ -1 & 3 & 2 \\ 0 & 1 & 4 \end{bmatrix} \begin{bmatrix} c_1 \\ c_2 \\ c_3 \end{bmatrix} = \begin{bmatrix} 0 \\ 0 \\ 0 \end{bmatrix}.$$

Next we reduce the coefficient matrix to row echelon form to determine the solution for (c_1, c_2, c_3).

$$\begin{bmatrix} 1 & -2 & 1 \\ -1 & 3 & 2 \\ 0 & 1 & 4 \end{bmatrix} \xrightarrow{R_2 + R_1} \begin{bmatrix} 1 & -2 & 1 \\ 0 & 1 & 3 \\ 0 & 1 & 4 \end{bmatrix} \xrightarrow{R_3 - R_2} \begin{bmatrix} 1 & -2 & 1 \\ 0 & 1 & 3 \\ 0 & 0 & 1 \end{bmatrix} \implies \begin{bmatrix} c_1 \\ c_2 \\ c_3 \end{bmatrix} = \begin{bmatrix} 0 \\ 0 \\ 0 \end{bmatrix}.$$

That is, the vector equation has only the trivial solution for the coefficients. Therefore W is linearly independent.

Now we use the Gram–Schmidt process to obtain an orthogonal basis:

$$v_1 = w_1 = (1, -1, 0);$$

$$v_2 = w_2 - \frac{v_1^T w_2}{v_1^T v_1} v_1 = w_2 - \frac{v_1 \cdot w_2}{\|v_1\|^2} v_1$$

$$= (-2, 3, 1) - \frac{(-5)(1, -1, 0)}{2} = \left(\frac{1}{2}, \frac{1}{2}, 1\right);$$

$$v_3 = w_3 - \frac{v_1^T w_3}{v_1^T v_1} v_1 - \frac{v_2^T w_3}{v_2^T v_2} v_2 = w_3 - \frac{v_1 \cdot w_3}{\|v_1\|^2} v_1 - \frac{v_2 \cdot w_3}{\|v_2\|^2} v_2$$

$$=(1, 2, 4) - \frac{(-1)(1, -1, 0)}{2} - \frac{\left(-\frac{11}{2}\right)\left(\frac{1}{2}, \frac{1}{2}, 1\right)}{3/2} = \left(-\frac{1}{3}, -\frac{1}{3}, \frac{1}{3}\right).$$

Then $\{v_1, v_2, v_3\}$ is an orthogonal basis of W.

ii) Since $\{v_1, v_2, v_3\}$ is an orthogonal basis, we have

$$u = \frac{v_1^T u}{v_1^T v_1} v_1 + \frac{v_2^T u}{v_2^T v_2} v_2 + \frac{v_3^T u}{v_3^T v_3} v_3.$$

Therefore, we have

$$[u]_V = \left(\frac{v_1^T u}{v_1^T v_1}, \frac{v_2^T u}{v_2^T v_2}, \frac{v_3^T u}{v_3^T v_3}\right) = \left(\frac{\langle u, v_1 \rangle}{\|v_1\|^2}, \frac{\langle u, v_2 \rangle}{\|v_2\|^2}, \frac{\langle u, v_3 \rangle}{\|v_3\|^2}\right) = (0, 2, 0).$$

$\square$

Remark 4.4.4. (QR **decomposition**) Let $A = [u_1 : u_2 : \cdots : u_m]$ be an $n \times m$ matrix with m linearly independent vectors. Then using the Gram–Schmidt process the columns of A can be converted into an orthonormal set of vectors which are columns of $Q = [q_1 : q_2 : \cdots : q_m]$. Then the transition matrix from basis $\{u_1, u_2, \cdots, u_m\}$ to $\{q_1, q_2, \cdots, q_m\}$ is

$$R = \begin{bmatrix} u_1^T q_1 & u_2^T q_1 & \cdots & u_m^T q_1 \\ u_1^T q_2 & u_2^T q_2 & \cdots & u_m^T q_2 \\ \vdots & \vdots & \ddots & \vdots \\ u_1^T q_m & u_2^T q_m & \cdots & u_m^T q_m \end{bmatrix}.$$

By the Gram–Schmidt process we know that $u_i^T q_j = 0$ for every $i < j$ because $q_j \perp \mathrm{span}\{u_1, u_2, \cdots, u_{j-1}\}$. Therefore R is a upper triangular matrix with

$$R = \begin{bmatrix} u_1^T q_1 & u_2^T q_1 & \cdots & u_m^T q_1 \\ 0 & u_2^T q_2 & \cdots & u_m^T q_2 \\ \vdots & \vdots & \ddots & \vdots \\ 0 & 0 & \cdots & u_m^T q_m \end{bmatrix}.$$

That is, $A = QR$ and A is decomposed into the product of an orthogonal matrix and an upper triangular matrix. $\square$

Example 4.4.5. Find the QR decomposition of the matrix

$$A = \begin{bmatrix} 0 & -1 & 4 \\ 1 & 1 & -5 \\ 1 & 0 & 0 \\ 0 & 1 & 0 \\ 0 & 0 & 1 \end{bmatrix} = [c_1 \ c_2 \ c_3],$$

where c_1, c_2 and c_3 are the columns of A.

Solution: The matrix contains a block of the identity matrix I_3. Therefore we have $\text{rank}(A) = 3$. Then the dimension of the column space of A is 3 and the columns of A form a basis. Applying the Gram-Schmidt process to the basis c_1, c_2 and c_3 of the column space we have

$$\beta_1 = c_1,$$

$$\beta_2 = c_2 - \frac{c_1^T \beta_1}{\beta_1^T \beta_1} \cdot \beta_1 = \frac{1}{2}(-2, 1, -1, 2, 0),$$

$$\beta_3 = c_3 - \frac{c_3^T \beta_2}{\beta_2^T \beta_2} \cdot \beta_2 - \frac{c_3^T \beta_1}{\beta_1^T \beta_1} \cdot \beta_1 = \frac{1}{5}(7, -6, 6, 13, 5).$$

Normalization on $\{\beta_1, \beta_2, \beta_3\}$ gives

$$q_1 = \frac{1}{\sqrt{2}}(0, 1, 1, 0, 0),$$

$$q_2 = \frac{1}{\sqrt{10}}(-2, 1, -1, 2, 0),$$

$$q_3 = \frac{1}{\sqrt{315}}(7, -6, 6, 13, 5).$$

Then the matrix Q is

$$Q = \begin{bmatrix} 0 & \frac{-2}{\sqrt{10}} & \frac{7}{\sqrt{315}} \\ \frac{1}{\sqrt{2}} & \frac{1}{\sqrt{10}} & \frac{-6}{\sqrt{315}} \\ \frac{1}{\sqrt{2}} & \frac{-1}{\sqrt{10}} & \frac{6}{\sqrt{315}} \\ 0 & \frac{2}{\sqrt{10}} & \frac{13}{\sqrt{315}} \\ 0 & 0 & \frac{5}{\sqrt{315}} \end{bmatrix}$$

and the matrix R is

$$R = \begin{bmatrix} c_1^T q_1 & c_2^T q_1 & c_3^T q_1 \\ 0 & c_2^T q_2 & c_3^T q_2 \\ 0 & 0 & c_3^T q_3 \end{bmatrix} = \begin{bmatrix} \sqrt{2} & \frac{\sqrt{2}}{2} & -\frac{5\sqrt{2}}{2} \\ 0 & \frac{\sqrt{10}}{2} & -\frac{13}{\sqrt{10}} \\ 0 & 0 & \frac{\sqrt{63}}{\sqrt{5}} \end{bmatrix}.$$

Therefore we have

$$A = \begin{bmatrix} 0 & -1 & 4 \\ 1 & 1 & -5 \\ 1 & 0 & 0 \\ 0 & 1 & 0 \\ 0 & 0 & 1 \end{bmatrix} = \begin{bmatrix} 0 & \frac{-2}{\sqrt{10}} & \frac{7}{\sqrt{315}} \\ \frac{1}{\sqrt{2}} & \frac{1}{\sqrt{10}} & \frac{-6}{\sqrt{315}} \\ \frac{1}{\sqrt{2}} & \frac{-1}{\sqrt{10}} & \frac{6}{\sqrt{315}} \\ 0 & \frac{2}{\sqrt{10}} & \frac{13}{\sqrt{315}} \\ 0 & 0 & \frac{5}{\sqrt{315}} \end{bmatrix} \begin{bmatrix} \sqrt{2} & \frac{\sqrt{2}}{2} & -\frac{5\sqrt{2}}{2} \\ 0 & \frac{\sqrt{10}}{2} & -\frac{13}{\sqrt{10}} \\ 0 & 0 & \frac{\sqrt{63}}{\sqrt{5}} \end{bmatrix}.$$

□

Exercise 4.4.6.

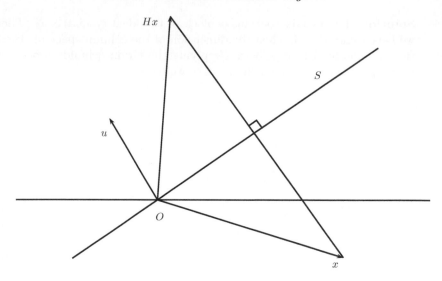

FIGURE 4.3: Hx is the mirror reflection of x with respect to the hyperplane S with unit normal u.

1. Let $W = \text{span}\{w_1, w_2, w_3\}$ be a subspace of $\mathbb{R}^4$, where

$$w_1 = (1, -1, 0, 1), w_2 = (2, -2, 3, 1), w_3 = (1, 2, 4, 0).$$

i) Show that $\{w_1, w_2, w_3\}$ is linearly independent. Then use the Gram-Schmidt process to convert $\{w_1, w_2, w_3\}$ into an orthogonal basis $V = \{v_1, v_2, v_3\}$ of W. ii) Find the coordinate vector $[u]_V$ of $u = (1, 0, 2, 0)$ relative to the orthogonal basis V.

2. Show that there exists an $m \times n$, $n < m$ matrix A such that $A^T A = I_n$ but $AA^T \neq I_m$, where I_n and I_m are $n \times n$ and $m \times m$ identity matrices, respectively.

3. Show that every finite dimensional inner product space has an orthonormal basis.

4. Let $A = \begin{bmatrix} -1 & 1 & 1 & 1 \\ 1 & -1 & 1 & 1 \\ 1 & 1 & -1 & 1 \\ 1 & 1 & 1 & -1 \end{bmatrix}$. Find a QR decomposition of A, if it exists.

5. Let $A = \begin{bmatrix} -1 & 1 \\ 1 & -1 \\ 1 & 1 \\ 1 & 1 \end{bmatrix}$. Find a QR decomposition of A, if it exists.

6. Let $A = [c_1 : c_2 : \cdots : c_m]$ be an $n \times m$ matrix with m linearly independent vectors. Suppose that the Gram-Schmidt process converted the columns of A into an orthogonal set of vectors which are columns of $B = [b_1 : b_2 : \cdots : b_m]$. Find the matrix R such that

$$A = BR.$$

7. Let $S = \{1, x, x^2\}$ be a basis for P_2. Define inner product on P_2 by

$$\langle f, g \rangle = \int_{-1}^{1} f(x)g(x)\mathrm{d}x.$$

If we replace the dot product in the Gram-Schmidt process with inner product, can you convert S into an orthogonal basis for P_2?

8. Let $u \in \mathbb{R}^n$ be a unit vector. What is the representation matrix H of the mirror reflection about the plane orthogonal to u? Is H an orthogonal matrix? See Figure 4.3.

9. Let v be a unit vector in $\mathbb{R}^n$. Show that the matrix $P = I - 2vv^T$ is an orthogonal and symmetric matrix. We call $P = I - 2vv^T$ with $v^T v = 1$ a **Householder matrix** and call the map $x \to Px$, $x \in \mathbb{R}^n$ a Householder transformation.

10. Let $x, y \in \mathbb{R}^n$ be such that $x \neq y$ and $\|x\| = \|y\|$. Let $u = \frac{x-y}{\|x-y\|}$ and $P = I - 2uu^T$.

i) Show that $Px = y$ and $Py = x$.

ii) Show that for every $x \in \mathbb{R}^n$, $y = \pm\|x\|e_1$, where e_1 is the first vector of the standard basis of $\mathbb{R}^n$, there exists a Householder matrix P such that $Px = y$.

11. Let $x = (3, 4, 0, 0) \in \mathbb{R}^4$ and $y = (5, 0, 0, 0) \in \mathbb{R}^4$. Find a Householder matrix such that $Px = y$.

12. Let $A = \begin{bmatrix} 1 & -1 & 0 \\ 1 & 1 & 0 \\ 0 & 1 & 1 \end{bmatrix}$. Use the Householder transformation to find an orthogonal matrix Q such that $Q^T A$ is upper triangular.

13. Let A be an $m \times n$ matrix with linearly independent columns. Show that there exists a sequence of orthogonal matrices $P_1, P_2, \cdots, P_m$ such that

$$P_m P_{m-1} \cdots P_2 P_1 A = R$$

is upper triangular.

14. Explain that a product of symmetric orthogonal matrices is orthogonal but may not be symmetric.

Chapter 5

Determinants

5.1 Introduction to determinants

We know that the 2×2 matrix $A = \begin{bmatrix} a & b \\ c & d \end{bmatrix}$ is invertible if and only if the scalar $ad - bc \neq 0$. And we also learned that elementary row operations do not change invertibility of a matrix. If we regard the scalar $ad - bc$ as the value of a function, called **determinant** det acting on the matrix A, we wish to know how elementary row operations will change the function value.

1) Multiply a row of A by a constant t. Then

$$\det\left(\begin{bmatrix} ta & tb \\ c & d \end{bmatrix}\right) = tad - tbc$$
$$= t(ad - bc)$$
$$= t\det(A).$$

Namely, det is a homogeneous function of its rows. Moreover, let $E_1 = \begin{bmatrix} t & 0 \\ 0 & 1 \end{bmatrix}$. We have

$$\det(E_1 A) = \det(E_1)\det(A).$$

2) Add a multiple of one row to another. Then

$$\det\left(\begin{bmatrix} a & b \\ ta + c & tb + d \end{bmatrix}\right) = a(tb+d) - b(ta+c) = tab + ad - tab - bc = \det(A).$$

That is, adding a multiple of one row to another does not change the determinant. Moreover, let $E_2 = \begin{bmatrix} 1 & 0 \\ t & 1 \end{bmatrix}$. We have

$$\det(E_2 A) = \det(E_2)\det(A).$$

3) Interchange two rows. Then

$$\det\left(\begin{bmatrix} c & d \\ a & b \end{bmatrix}\right) = bc - ad = -\det(A).$$

That is, interchanging two rows changes the sign of the determinant. Let $E_3 = \begin{bmatrix} 0 & 1 \\ 1 & 0 \end{bmatrix}$. We have

$$\det(E_3 A) = \det(E_3)\det(A).$$

Items 1) and 2) hinted that det may be a linear function of the rows of a matrix. Indeed,

$$\det\left(\begin{bmatrix} a+e & b+f \\ c & d \end{bmatrix}\right) = d(a+e) - c(b+f) = ad - bc + ed - fc$$

$$= \det\left(\begin{bmatrix} a & b \\ c & d \end{bmatrix}\right) + \det\left(\begin{bmatrix} e & f \\ c & d \end{bmatrix}\right)$$

$$\det\left(\begin{bmatrix} ta & tb \\ c & d \end{bmatrix}\right) = t\det(A).$$

Let M_{nn} be the vector space of $n \times n$ matrices. We want to define a function $\det : M_{nn} \to \mathbb{R}$ called a **determinant** which satisfies the following properties.

P1) det is linear in each of the rows of $A \in M_{nn}$. For every $A = \begin{bmatrix} a_1 \\ a_2 \\ \vdots \\ a_n \end{bmatrix}$ in M_{nn}, $f \in \mathbb{R}^n$, $t \in \mathbb{R}$,

$$\det\begin{bmatrix} a_1 \\ a_2 \\ \vdots \\ a_i+f \\ \vdots \\ a_n \end{bmatrix} = \det\begin{bmatrix} a_1 \\ a_2 \\ \vdots \\ a_i \\ \vdots \\ a_n \end{bmatrix} + \det\begin{bmatrix} a_1 \\ a_2 \\ \vdots \\ f \\ \vdots \\ a_n \end{bmatrix}, \quad \det\begin{bmatrix} a_1 \\ a_2 \\ \vdots \\ ta_i \\ \vdots \\ a_n \end{bmatrix} = t\det\begin{bmatrix} a_1 \\ a_2 \\ \vdots \\ a_i \\ \vdots \\ a_n \end{bmatrix}.$$

P2) det changes sign if two rows of $A \in M_{nn}$ are interchanged:

$$\det\begin{bmatrix} a_1 \\ a_2 \\ \vdots \\ a_j \\ \vdots \\ a_i \\ \vdots \\ a_n \end{bmatrix} = -\det\begin{bmatrix} a_1 \\ a_2 \\ \vdots \\ a_i \\ \vdots \\ a_j \\ \vdots \\ a_n \end{bmatrix}$$

P3) Let I be the identity matrix.

$$\det(I) = 1.$$

Determinant of permutation matrices

A permutation matrix P is a matrix obtained by interchanging rows of the identity matrix I, and we know that $P^T = P^{-1}$. By property P2) and P3), we know that

$$\det(P) = \pm 1.$$

The sign of $\det(P)$ is determined by the number of interchangings (or called **transpositions**) of the rows of I in order to obtain P. If the number is odd, $\det(P) = -1$; if it is even, $\det(P) = 1$. To determine the number of transpositions, we assume that the rows of P are a permutation of the row of I, which is an ordering of the numbers $\{1, 2, \cdots, n\}$. Let S denote the set of all permutations of the set of numbers $\{1, 2, \cdots, n\}$. Every member $\sigma = (\sigma(1), \sigma(2), \cdots, \sigma(n)) \in S$ can be represented as

$$\sigma = \begin{pmatrix} 1 & 2 & \cdots & n \\ \sigma(1) & \sigma(2) & \cdots & \sigma(n) \end{pmatrix},$$

where the first row of σ denotes the row number of I and the value of $\sigma(i)$ indicates that the current i-th row of P was the $\sigma(i)$-th row of I. For example, the following representation of the permutation σ,

$$\sigma = \begin{pmatrix} 1 & 2 & 3 & 4 & 5 \\ 5 & 3 & 2 & 1 & 4 \end{pmatrix},$$

implies that the current first row of P was the fifth row of I, the current second row of P was the third row of I, the current third row of P was the second row of I, and so on.

Now we count how many transpositions to achieve σ from the $(12 \cdots n)$. We use the so-called total **inversions** $\tau(\sigma)$ in σ which is the cardinality of all the occurrences that $\sigma(j) < \sigma(i)$ with $j > i$. That is,

$$\tau(\sigma) = \sum_{i=1}^{n} \text{number of } (i, j)\text{'s with } \sigma(i) > \sigma(j) \text{ and } i < j.$$

For example $\tau(53214) = 4+2+1+0+0 = 7$, $\tau(13542) = 0+1+2+1+0 = 4$. It turns out that $\tau(\sigma)$ is the minimal number of transpositions to achieve σ from the $(12 \cdots n)$. Then by Property P3), we have

---

Lemma 5.1.1.
$$\det(P) = (-1)^{\tau(\sigma)},$$

where σ is the permutation of the rows of I to obtain the permutation matrix P and $\tau(\sigma)$ is the total inversions of σ.

Example 5.1.2.

1) Let

$$P = \begin{bmatrix} 0 & 1 & 0 \\ 1 & 0 & 0 \\ 0 & 0 & 1 \end{bmatrix}.$$

Then P is a permutation matrix obtained from I by the permutation $\sigma = (213)$ with representation $\sigma = \begin{pmatrix} 1 & 2 & 3 \\ 2 & 1 & 3 \end{pmatrix}$ and with $\tau(\sigma) = \tau(213) = 1 + 0 + 0 = 1$. Then $\det(P) = (-1)^1 = -1$.

2) Let

$$P = \begin{bmatrix} 0 & 1 & 0 & 0 & 0 \\ 0 & 0 & 0 & 1 & 0 \\ 1 & 0 & 0 & 0 & 0 \\ 0 & 0 & 0 & 0 & 1 \\ 0 & 0 & 1 & 0 & 0 \end{bmatrix}.$$

Then P is a permutation matrix obtained from I by the permutation $\sigma = (24153)$ with representation $\sigma = \begin{pmatrix} 1 & 2 & 3 & 4 & 5 \\ 2 & 4 & 1 & 5 & 3 \end{pmatrix}$ and with $\tau(\sigma) = \tau(24153) = 1 + 2 + 0 + 1 + 0 = 4$. Then $\det(P) = (-1)^4 = 1$.   □

Before we discuss determinants of general $n \times n$ matrices, let us count the inversions of the inverse σ^{-1} of a permutation $\sigma \in S$, which is the permutation which restores the permutation σ into identity id. That is, $\sigma^{-1} \circ \sigma = id$. For example,

$$\sigma = \begin{pmatrix} 1 & 2 & 3 & 4 & 5 \\ 5 & 3 & 2 & 1 & 4 \end{pmatrix}, \ \sigma^{-1} = \begin{pmatrix} 1 & 2 & 3 & 4 & 5 \\ \sigma^{-1}(1) & \sigma^{-1}(2) & \sigma^{-1}(3) & \sigma^{-1}(4) & \sigma^{-1}(5) \end{pmatrix}$$

$$= \begin{pmatrix} 1 & 2 & 3 & 4 & 5 \\ 4 & 3 & 2 & 5 & 1 \end{pmatrix}.$$

By examining the matrix representation of permutations, we have

$$\begin{aligned} \tau(\sigma) &= \text{number of } \{(i, j) : \sigma(i) > \sigma(j) \text{ with } i < j\} \\ &= \text{number of } \{(\sigma^{-1}(s), \sigma^{-1}(t)) : s > t \text{ with } \sigma^{-1}(s) < \sigma^{-1}(t)\} \\ &= \text{number of } \{(s, t) : s > t \text{ with } \sigma^{-1}(s) < \sigma^{-1}(t)\} \\ &= \tau(\sigma^{-1}). \end{aligned}$$

In summary we have

> **Lemma 5.1.3.** Let σ be a permutation of the set $\{1, 2, \cdots, n\}$. Then we have
> $$\tau(\sigma) = \tau(\sigma^{-1}),$$
> where $\tau(\sigma)$ is the total of the inversions of σ.

Determinants of $n \times n$ matrices

Let $A = (a_{ij})$ be an $n \times n$ matrix with rows $a_1, a_2, \cdots, a_n$. Let $\{e_1, e_2, \cdots, e_n\}$ be the standard basis of $\mathbb{R}^n$. Then we have

$$
\begin{aligned}
a_1 &= a_{11}e_1^T + a_{12}e_2^T + \cdots + a_{1n}e_n^T, \\
a_2 &= a_{21}e_1^T + a_{22}e_2^T + \cdots + a_{2n}e_n^T, \\
&\;\;\vdots \qquad\qquad \vdots \\
a_n &= a_{n1}e_1^T + a_{n2}e_2^T + \cdots + a_{nn}e_n^T.
\end{aligned}
$$

By linearity property of determinants, we have

$$
\det(A) = \det\left(\begin{bmatrix} a_{11}e_1^T \\ a_2 \\ \vdots \\ a_n \end{bmatrix}\right) + \det\left(\begin{bmatrix} a_{12}e_2^T \\ a_2 \\ \vdots \\ a_n \end{bmatrix}\right) + \cdots + \det\left(\begin{bmatrix} a_{1n}e_n^T \\ a_2 \\ \vdots \\ a_n \end{bmatrix}\right)
$$

$$
= a_{11}\det\left(\begin{bmatrix} e_1^T \\ a_2 \\ \vdots \\ a_n \end{bmatrix}\right) + a_{12}\det\left(\begin{bmatrix} e_2^T \\ a_2 \\ \vdots \\ a_n \end{bmatrix}\right) + \cdots + a_{1n}\det\left(\begin{bmatrix} e_n^T \\ a_2 \\ \vdots \\ a_n \end{bmatrix}\right)
$$

$$
= \sum_{k_1=1}^{n} a_{1k_1}\det\left(\begin{bmatrix} e_{k_1}^T \\ a_2 \\ \vdots \\ a_n \end{bmatrix}\right)
$$

$$
= \sum_{k_1=1}^{n} a_{1k_1}\left(\sum_{k_2=1}^{n} a_{2k_2}\det\left(\begin{bmatrix} e_{k_1}^T \\ e_{k_2}^T \\ \vdots \\ a_n \end{bmatrix}\right)\right)
$$

$$
= \sum_{k_1=1}^{n} a_{1k_1} \sum_{k_2=1}^{n} a_{2k_2} \cdots \sum_{k_n=1}^{n} a_{nk_n}\det\left(\begin{bmatrix} e_{k_1}^T \\ e_{k_2}^T \\ \vdots \\ e_{k_n}^T \end{bmatrix}\right)
$$

$$
= \sum_{k_1=1}^{n} \sum_{k_2=1}^{n} \cdots \sum_{k_n=1}^{n} a_{1k_1}a_{2k_2}\cdots a_{nk_n}\det\begin{bmatrix} e_{k_1}^T \\ e_{k_2}^T \\ \vdots \\ e_{k_n}^T \end{bmatrix}
$$

$$
= \sum_{\sigma \in S} a_{1\sigma(1)}a_{2\sigma(2)} \cdots a_{n\sigma(n)}(-1)^{\tau(\sigma)},
$$

where the last step is based on the observation that, by P2),

$$\det \begin{bmatrix} e_{k_1}^T \\ e_{k_2}^T \\ \vdots \\ e_{k_n}^T \end{bmatrix} = 0, \quad \text{if } k_i = k_j,$$

since switching the identical two rows will not change the matrix but will reverse the sign of the determinant. Therefore only the permutation matrices survive to have nonzero determinants in the summation. That is, the determinant of an $n \times n$ matrix is the sum of every such signed product of n entries of A that contains a unique entry from each row and each column. Therefore, we have

Theorem 5.1.4. If $\det : M_{nn} \to \mathbb{R}$ satisfies properties P1), P2) and P3), then its value is uniquely determined by

$$\det(A) = \sum_{\sigma \in S} (-1)^{\tau(\sigma)} a_{1\sigma(1)} a_{2\sigma(2)} \cdots a_{n\sigma(n)},$$

where S is the set of all permutations of the set $\{1, 2, \cdots, n\}$ and $\tau(\sigma)$ is the inversion of σ.

An immediate consequence is that

Corollary 5.1.5.

i) If $A \in M_{nn}$ has a zero row or zero column, then $\det(A) = 0$.

ii) If $A = (a_{ij})$ is a triangular matrix, then $\det(A) = a_{11} a_{22} \cdots a_{nn}$.

We leave it as an exercise to show that the function $f : M_{nn} \to \mathbb{R}$ defined by

$$f(A) = \sum_{\sigma \in S} (-1)^{\tau(\sigma)} a_{1\sigma(1)} a_{2\sigma(2)} \cdots a_{n\sigma(n)}$$

must satisfy properties P1), P2) and P3).

Exercise 5.1.6.

1. Let A be an $n \times n$ matrix. If A has two proportional rows, then $\det(A) = 0$.

2. Let S be the set of all permutations of the numbers $(1, 2, 3, \cdots, n)$. We call a permutation an even (odd) permutation if its total inversion is even (odd, respectively). Show that there are the same number of odd permutations and even permutations.

3. Let S be the set of all permutations of the numbers $(1, 2, 3, \cdots, n)$ and $\sigma \in S$. Show that every transposition on σ changes its parity.

4. Find the total inversion, parity and matrix representation of each of the following permutations

    i)   (123564);

    ii) (456213);

    iii) (123456).

5. Let A be a 5×5 matrix. Find the coefficient of each the following products in the permutation expansion of $\det(A)$.

    i) $a_{15}a_{24}a_{33}a_{42}a_{51}$;

    ii) $a_{13}a_{25}a_{31}a_{42}a_{54}$;

    iii) $a_{14}a_{23}a_{32}a_{45}a_{51}$;

    iv) $a_{12}a_{23}a_{34}a_{45}a_{51}$.

6. Find the coefficients of the terms of x^3 and x^4 for the polynomial f defined by

$$f(x) = \det \begin{bmatrix} x & 2 & x & 1 \\ 1 & 3x & 1 & -1 \\ 2 & 1 & x & -2 \\ 3 & 1 & 2 & x \end{bmatrix}.$$

7. Find the determinants of the following matrices.

$$A = \begin{bmatrix} 0 & 1 & 0 & \cdots & 0 \\ 0 & 0 & 2 & \cdots & 0 \\ \vdots & \vdots & \vdots & \ddots & 0 \\ 0 & 0 & 0 & \cdots & n-1 \\ n & 0 & 0 & \cdots & 0 \end{bmatrix}, \quad B = \begin{bmatrix} 0 & 0 & 1 & 0 & \cdots & 0 \\ 0 & 0 & 0 & 2 & \cdots & 0 \\ \vdots & \vdots & \vdots & \vdots & \ddots & n-2 \\ n-1 & 0 & 0 & 0 & \cdots & 0 \\ 0 & n & 0 & 0 & \cdots & 0 \end{bmatrix}.$$

8. Let A be an $n \times n$ matrix such that

$$a_{ij} = \begin{cases} 1 & \text{if } i+j = n+1 \\ 0 & \text{otherwise.} \end{cases}$$

Find $\det(A)$.

9. Let A be an $n \times n$ matrix. Show that if A has more than $n^2 - n$ zero entries, then $\det(A) = 0$.

10. Show that the function $f : M_{nn} \to \mathbb{R}$ defined by

$$f(A) = \sum_{\sigma \in S} (-1)^{\tau(\sigma)} a_{1\sigma(1)} a_{2\sigma(2)} \cdots a_{n\sigma(n)}$$

must satisfy properties P1), P2) and P3).

5.2 Properties of determinants

We know that for 2×2 matrix A, $\det(A) = \det(A^T)$. Indeed it is valid for $A \in M_{nn}$.

> **Theorem 5.2.1.** $\det(A) = \det(A^T)$.

Proof. We have

$$\det(A) = \sum_{\sigma \in S} (-1)^{\tau(\sigma)} a_{1\sigma(1)} a_{2\sigma(2)} \cdots a_{n\sigma(n)}.$$

Consider $a_{1\sigma(1)} a_{2\sigma(2)} \cdots a_{n\sigma(n)}$, where $\sigma \in S$ is the column permutation with the row permutation fixed with $(12 \cdots n)$. If the column permutation is fixed with $(12 \cdots n)$, the row permutation is exactly σ^{-1}. By Lemma 5.1.3, we have $\tau(\sigma) = \tau(\sigma^{-1})$. Namely, the total of the inversions created by σ is restored by σ^{-1} to 0. Hence $\tau(\sigma) = \tau(\sigma^{-1})$. Therefore, one can fix the column permutation to $(12 \cdots n)$ and obtain that

$$\det(A) = \sum_{\sigma \in S} (-1)^{\tau(\sigma)} a_{\sigma^{-1}(1)\,1} a_{\sigma^{-1}(2)\,2} \cdots a_{\sigma^{-1}(n)\,n}$$

$$= \sum_{\sigma \in S} (-1)^{\tau(\sigma^{-1})} a^T_{1,\,\sigma^{-1}(1)} a^T_{2,\,\sigma^{-1}(2)} \cdots a^T_{n,\,\sigma^{-1}(n)}$$

$$= \sum_{\sigma^{-1} \in S} (-1)^{\tau(\sigma^{-1})} a^T_{1,\,\sigma^{-1}(1)} a^T_{2,\,\sigma^{-1}(2)} \cdots a^T_{n,\,\sigma^{-1}(n)}$$

$$= \det(A^T).$$

$\square$

A significant consequence of Theorem 5.2.1 is that the properties of determinants described at P1) and P2) with respect to rows are also valid in terms of columns. Now we use Theorem 5.2.1 to establish the co-factor formula which can be used to compute determinants.

Let $A = (a_{ij})$ be an $n \times n$ matrix with rows $a_1, a_2, \cdots, a_n$. Let M_{ij} denote the submatrix of A with the i-th row and the j-th column deleted. We have

$$a_i = a_{i1} e_1^T + a_{i2} e_2^T + \cdots + a_{in} e_n^T$$

and

$$\det(A) = \det\left(\begin{bmatrix} a_1 \\ \vdots \\ a_{i1} e_1^T \\ \vdots \\ a_n \end{bmatrix}\right) + \det\left(\begin{bmatrix} a_1 \\ \vdots \\ a_{i2} e_2^T \\ \vdots \\ a_n \end{bmatrix}\right) + \cdots + \det\left(\begin{bmatrix} a_1 \\ \vdots \\ a_{in} e_n^T \\ \vdots \\ a_n \end{bmatrix}\right)$$

$$=a_{i1} \det \left(\begin{bmatrix} a_1 \\ \vdots \\ e_1^T \\ \vdots \\ a_n \end{bmatrix} \right) + a_{i2} \det \left(\begin{bmatrix} a_1 \\ \vdots \\ e_2^T \\ \vdots \\ a_n \end{bmatrix} \right) + \cdots + a_{in} \det \left(\begin{bmatrix} a_1 \\ \vdots \\ e_n^T \\ \vdots \\ a_n \end{bmatrix} \right).$$

By means of interchanging rows and columns, we can manage to put the entry 1 of e_j^T from row i into the (n, n)-position, using $(n-i)+(n-j)$ interchangings. Then by property 2), and noticing $(-1)^{n-i+n-j} = (-1)^{i+j}$, we have

$$\det(A) = a_{i1}(-1)^{i+1} \det \begin{bmatrix} M_{i1} & * \\ \mathbf{0} & 1 \end{bmatrix} + \cdots + a_{ij}(-1)^{i+j} \det \begin{bmatrix} M_{ij} & * \\ \mathbf{0} & 1 \end{bmatrix}$$

$$+ \cdots + a_{in}(-1)^{i+n} \det \begin{bmatrix} M_{in} & * \\ \mathbf{0} & 1 \end{bmatrix}$$

$$= a_{i1}(-1)^{i+1} \det M_{i1} + \cdots + a_{ij}(-1)^{i+j} \det M_{ij}$$

$$+ \cdots + a_{in}(-1)^{i+n} \det M_{in}.$$

Theorem 5.2.2. (co-factor expansion) Let $A = (a_{ij})$ be an $n \times n$ matrix and M_{ij} be the submatrix of A after the i-th row and the j-th column of A are deleted. Then

$$\det(A) = a_{i1}(-1)^{i+1} \det M_{i1} + \cdots + a_{ij}(-1)^{i+j} \det M_{ij}$$

$$+ \cdots + a_{in}(-1)^{i+n} \det M_{in}.$$

Define $C_{ij} = (-1)^{i+j} \det M_{ij}$. We call C_{ij} the **co-factor** associated with the entry a_{ij}. We leave it as an exercise to show that co-factor expansion is also valid in terms of columns.

Example 5.2.3. Let $A = \begin{bmatrix} 1 & 2 & 3 \\ 4 & 5 & 6 \\ 7 & 8 & 9 \end{bmatrix}$. By the co-factor expansion formula, we have

$$\det(A) = 4 \cdot (-1)^{2+1} \det \begin{bmatrix} 2 & 3 \\ 8 & 9 \end{bmatrix} + 5 \cdot (-1)^{2+2} \det \begin{bmatrix} 1 & 3 \\ 7 & 9 \end{bmatrix}$$

$$+ 6 \cdot (-1)^{2+3} \det \begin{bmatrix} 1 & 2 \\ 7 & 8 \end{bmatrix}$$

$$= (-4) \cdot (-6) + 5 \cdot (-12) + (-6) \cdot (-6)$$

$$= 0.$$

In principle, one can use the co-factor formula to compute the determinant of every $n \times n$ matrix, by recursively reducing the sizes of the submatrices involved in the co-factor formula. However, the amount of computation is huge when n is large. In fact, elementary row operations also play important roles in the computation of determinants.

Theorem 5.2.4. Let $A = (a_{ij})$ be an $n \times n$ matrix. If E is an elementary matrix, then

$$\det(EA) = \det(E)\det(A).$$

Proof. We distinguish three cases:

i) If E denotes the row operation of interchanging two rows, say, row i interchanged with row j ($i > j$), then E is a permutation matrix which is obtained by interchanging row i with row j of I, involving $2(j - i) - 1$ transpositions. Therefore

$$\det(E) = (-1)^{2(j-i)-1} = -1.$$

On the other hand, by Property 2), we have $\det(EA) = -\det(A) = (-1)\det(A)$. That is, $\det(EA) = \det(E)\det(A)$.

ii) If E denotes the row operation of adding r-multiple of one row to another, then E is a triangular matrix with 1 in every main diagonal position. By Corollary 5.1.5, we have $\det(E) = 1$. On the other hand, by Property 1), we have

$$\det(EA) = \det \begin{bmatrix} \vdots \\ a_i + ra_j \\ \vdots \end{bmatrix} = \det \begin{bmatrix} \vdots \\ a_i \\ \vdots \end{bmatrix} + r \det \begin{bmatrix} \vdots \\ a_j \\ \vdots \end{bmatrix},$$

where the displayed a_j is in the i-th row. Then $\det \begin{bmatrix} \vdots \\ a_j \\ \vdots \end{bmatrix} = 0$ since interchanging its i-th row and the j-th row does not change the matrix, but by Property 3), the determinant has to change sign. Then we have $\det(EA) = \det(A)$. That is, $\det(EA) = \det(E)\det(A)$.

iii) If E denotes the row operation of multiplying a row by a constant $c \neq 0$, say, row i multiplied by c, then E is a diagonal matrix with 1 in every main diagonal position except for the c at the (i, i)-position. Then by Corollary 5.1.5, $\det(E) = c$. On the other hand, by Property 2), we have $\det(EA) = c \det(A)$. That is, $\det(EA) = \det(E)\det(A)$. □

Corollary 5.2.5. Let $A = (a_{ij})$ be an $n \times n$ matrix. If two rows of A are proportional, then

$$\det(A) = 0.$$

Theorem 5.2.6. A is invertible if and only if $\det(A) \neq 0$.

Proof. "$\Longrightarrow$" If A is invertible, then by Theorem 2.3.4, A is a product of elementary matrices. That is, there exist elementary matrices $E_1, E_2, \cdots, E_k$ such that $A = E_k E_{k-1} \cdots E_1$. By Theorem 5.2.7, we have $\det(A) = \det(E_k)) \det(E_{k-1}) \cdots \det(E_1)$ which is nonzero because the determinant of every elementary matrix is nonzero.

"$\Longleftarrow$" Suppose not. That is, A is not invertible, then by Theorem 2.3.4 $Ax = 0$ has nontrivial solutions. Hence the reduced row echelon form U of A has at least one zero row. That is, there exist elementary matrices $E_1, E_2, \cdots, E_k$ such that $E_k E_{k-1} \cdots E_1 A = U$. We have

$$\det(E_k E_{k-1} \cdots E_1 A) = \det(E_k) \det(E_{k-1}) \cdots \det(E_1) \det(A) = \det(U) = 0,$$

which lead to $\det(A) = 0$ since $\det(E_i) \neq 0$, $i = 1, 2, \cdots, k$. This is a contradiction. □

Theorem 5.2.7. Let A and B be $n \times n$ matrices. Then

$$\det(AB) = \det(A) \det(B).$$

Proof. If A is not invertible, then by Theorem 2.3.6, AB is not invertible. By Theorem 5.2.7, $\det(AB) = 0$ and $\det(A) = 0$. That is, $\det(AB) = \det(A) \det(B)$.

If A is invertible, then by Theorem 2.3.4, A is a product of elementary matrices. That is, there exist elementary matrices $E_1, E_2, \cdots, E_k$ such that $A = E_k E_{k-1} \cdots E_1$. By Theorem 5.2.7, we have

$$\begin{aligned} \det(AB) &= \det(E_k E_{k-1} \cdots E_1 B) \\ &= \det(E_k) \det(E_{k-1}) \cdots \det(E_1) \det(B) \\ &= \det(A) \det(B). \end{aligned}$$

□

Example 5.2.8. Compute the determinants of the following matrices

$$a) \quad A = \begin{bmatrix} 46159 & 46059 \\ 70281 & 70181 \end{bmatrix};$$

$$b) \quad B = \begin{bmatrix} 4 & 2 & 2 & 2 \\ 2 & 4 & 2 & 2 \\ 2 & 2 & 4 & 2 \\ 2 & 2 & 2 & 4 \end{bmatrix}.$$

Solution: a) Subtracting the second column from the first column we have

$$\det(A) = \det \begin{bmatrix} 100 & 46059 \\ 100 & 70181 \end{bmatrix} = 100(70181 - 46059) = 2412200.$$

b) Adding the second, the third and the fourth row to the first row, we get a row of 10's without changing the determinant.

$$\det(B) = \det \begin{bmatrix} 10 & 10 & 10 & 10 \\ 2 & 4 & 2 & 2 \\ 2 & 2 & 4 & 2 \\ 2 & 2 & 2 & 4 \end{bmatrix} \qquad \text{(Add row 2, 3, 4 to row 1)}$$

$$= 5 \cdot \det \begin{bmatrix} 2 & 2 & 2 & 2 \\ 2 & 4 & 2 & 2 \\ 2 & 2 & 4 & 2 \\ 2 & 2 & 2 & 4 \end{bmatrix} \qquad \text{(Factor 5 from row 1)}$$

$$= 5 \cdot \det \begin{bmatrix} 2 & 2 & 2 & 2 \\ 0 & 2 & 0 & 0 \\ 0 & 0 & 2 & 0 \\ 0 & 0 & 0 & 2 \end{bmatrix} \qquad \text{(Subtract row 1 from row 2, 3, 4)}$$

$$= 5 \cdot 2 \cdot 2 \cdot 2 \cdot 2 \qquad \text{(Determinant of a triangular matrix)}$$

$$= 80.$$

Since $AA^{-1} = I$, by Theorem 5.2.7 we have

Corollary 5.2.9. If A is an invertible matrix, then

$$\det(A^{-1}) = \frac{1}{\det(A)}.$$

Another application of Theorem 5.2.7 is the Cramer's rule. Consider $Ax = b$, where A is an $n \times n$ invertible matrix and $x, b \in \mathbb{R}^n$. Let $\{e_1, e_2, \cdots, e_n\}$ be the standard basis of $\mathbb{R}^n$. Since $Ax = b$ is consistent, the following matrix product holds:

$$A[e_1 : \cdots : e_{i-1} : x : e_{i+1} : \cdots : e_n] = [a_1 : \cdots : a_{i-1} : b : a_{i+1} : \cdots : a_n]. \quad (5.1)$$

Since the i-th coordinate of x is the only nontrivial entry in the i-th row of $[e_1 : \cdots : e_{i-1} : x : e_{i+1} : \cdots : e_n]$, we have by co-factor expansion

$$\det[e_1 : \cdots : e_{i-1} : x : e_{i+1} : \cdots : e_n] = x_i.$$

Let B_i denote the matrix obtained by replacing the i-th column of A with b. Equality (5.1) can be rewritten as

$$A[e_1 : \cdots : e_{i-1} : x : e_{i+1} : \cdots : e_n] = B_i,$$

which by Theorem 5.2.7 leads to

$$\det(A)x_i = \det(B_i).$$

In summary, we have

Corollary 5.2.10. (Cramer's rule) If A is an invertible matrix, then the solution of $Ax = b$ for $x = (x_1, x_2, \cdots, x_n)$ is

$$x_1 = \frac{\det(B_1)}{\det(A)}, \; x_2 = \frac{\det(B_2)}{\det(A)}, \; \cdots, \; x_n = \frac{\det(B_n)}{A},$$

where B_i, $i = 1, 2, \cdots, n$, is the matrix obtained by replacing the i-th column of A with b.

Example 5.2.11. Solve the following system of equations using Cramer's rule.

$$\begin{cases} 2x - y + 3z = 0 \\ x + 4y + 2z = 1 \\ 3x + 2y + z = 2. \end{cases}$$

Solution: The system of linear equations can be written as

$$\begin{bmatrix} 2 & -1 & 3 \\ 1 & 4 & 2 \\ 3 & 2 & 1 \end{bmatrix} \begin{bmatrix} x \\ y \\ z \end{bmatrix} = \begin{bmatrix} 0 \\ 1 \\ 2 \end{bmatrix}.$$

We note that

$$\det \begin{bmatrix} 2 & -1 & 3 \\ 1 & 4 & 2 \\ 3 & 2 & 1 \end{bmatrix} = -35 \neq 0.$$

Therefore, Cramer's rule applies. By Cramer's rule we have

$$x = \frac{\det \begin{bmatrix} \mathbf{0} & -1 & 3 \\ \mathbf{1} & 4 & 2 \\ \mathbf{2} & 2 & 1 \end{bmatrix}}{\det \begin{bmatrix} 2 & -1 & 3 \\ 1 & 4 & 2 \\ 3 & 2 & 1 \end{bmatrix}} = \frac{-21}{-35}, \; y = \frac{\det \begin{bmatrix} 2 & \mathbf{0} & 3 \\ 1 & \mathbf{1} & 2 \\ 3 & \mathbf{2} & 1 \end{bmatrix}}{\det \begin{bmatrix} 2 & -1 & 3 \\ 1 & 4 & 2 \\ 3 & 2 & 1 \end{bmatrix}} = \frac{-9}{-35}$$

and

$$z = \frac{\det \begin{bmatrix} 2 & -1 & \mathbf{0} \\ 1 & 4 & \mathbf{1} \\ 3 & 2 & \mathbf{2} \end{bmatrix}}{\det \begin{bmatrix} 2 & -1 & 3 \\ 1 & 4 & 2 \\ 3 & 2 & 1 \end{bmatrix}} = \frac{11}{-35}.$$

That is, the solution is

$$(x,\, y,\, z) = \left(\frac{21}{35},\, \frac{9}{35},\, -\frac{11}{35}\right).$$

We close this section with an application of the co-factor formula and Theorem 5.2.7. Consider $A \in M_{nn}$ with rows $a_1, a_2, \cdots, a_n, \cdots, n$. We have for every $i = 1, 2, \cdots, n$,

$$\det(A) = a_{i1}C_{i1} + a_{i2}C_{i2} + \cdots + a_{in}C_{in}.$$

It is interesting to examine $a_{j1}C_{i1} + a_{j2}C_{i2} + \cdots + a_{jn}C_{in}$ as well. It turns out that $a_{j1}C_{i1} + a_{j2}C_{i2} + \cdots + a_{jn}C_{in}$ is a cofactor expansion of the determinant of such a matrix that is NOT A, but is the matrix obtained by replacing the i-th row of A with its j-th row. Such a matrix has determinant zero because it has two identical rows. Therefore we have

$$a_{j1}C_{i1} + a_{j2}C_{i2} + \cdots + a_{jn}C_{in} = \begin{cases} \det(A) & \text{if } i = j, \\ 0 & \text{if } i \neq j \end{cases}$$

or equivalently

$$A \begin{bmatrix} C_{11} & C_{21} & \cdots & C_{n1} \\ C_{12} & C_{22} & \cdots & C_{n2} \\ \vdots & \vdots & \ddots & \vdots \\ C_{1n} & C_{2n} & \cdots & C_{nn} \end{bmatrix} = \det(A)I.$$

Definition 5.2.12. We call the transpose of the matrix of cofactors of A the adjoint of A, denoted by $\mathrm{adj}(A)$.

Theorem 5.2.13. Let A be an $n \times n$ matrix. Then

$$A \,\mathrm{adj}(A) = \det(A)I.$$

In particular, if A is invertible, then

$$A^{-1} = \frac{\mathrm{adj}(A)}{\det(A)}.$$

Exercise 5.2.14.

1. Let

$$A = \begin{bmatrix} 6 & 7 & 2 \\ 1 & 5 & 9 \\ 8 & 3 & 4 \end{bmatrix}.$$

Find $\det(A)$ by the following three methods: 1) Co-factor expansion; 2) Elementary row operations; 3) The permutation formula for determinants.

2. Find the determinants of the following matrices.

1) $A = \begin{bmatrix} 1 & 3 & 5 \\ 3 & 5 & 7 \\ 5 & 7 & 9 \end{bmatrix}$;   2) $B = \begin{bmatrix} 6 & 7 & 2 \\ 1 & 5 & 9 \\ 8 & 3 & 4 \end{bmatrix}$;

3) $C = \begin{bmatrix} 3 & 0 & 0 \\ 1 & 5 & 0 \\ -2 & -1 & 7 \end{bmatrix}$;   4) $D = \begin{bmatrix} 0 & 0 & 2 \\ 0 & 5 & 9 \\ 8 & 3 & 4 \end{bmatrix}$.

3. Find the determinants of the following matrices.

1) $A = \begin{bmatrix} 1 & 3 & 5 & 7 \\ 3 & 5 & 7 & 9 \\ 5 & 7 & 9 & 11 \\ 7 & 9 & 11 & 13 \end{bmatrix}$;   2) $B = \begin{bmatrix} 0 & 7 & 2 & 0 \\ 1 & 5 & 9 & 5 \\ 8 & 3 & 4 & 0 \\ -1 & 2 & 5 & 0 \end{bmatrix}$;

3) $C = \begin{bmatrix} 0 & 0 & 2 & 0 \\ 1 & 5 & 0 & 5 \\ 8 & 0 & 4 & 0 \\ 0 & 2 & 5 & 0 \end{bmatrix}$,   4) $D = \begin{bmatrix} 0 & 0 & 0 & 2 \\ 0 & 0 & 2 & 5 \\ 0 & 2 & 4 & 0 \\ 4 & 2 & 5 & 0 \end{bmatrix}$.

4. Find the determinants of the following matrices.

1) $A = \begin{bmatrix} 1 & 3 & 5 & 7 \\ 3 & 5 & 7 & 1 \\ 5 & 7 & 3 & 1 \\ 7 & 1 & 5 & 3 \end{bmatrix}$;   2) $B = \begin{bmatrix} 5 & 1 & 1 & 1 \\ 1 & 5 & 1 & 1 \\ 1 & 1 & 5 & 1 \\ 1 & 1 & 1 & 5 \end{bmatrix}$;

3) $C = \begin{bmatrix} a & b & c & d \\ b & a & d & c \\ c & d & a & b \\ d & c & b & a \end{bmatrix}$;   4) $D = \begin{bmatrix} x & y & y & y \\ y & x & y & y \\ y & y & x & y \\ y & y & y & x \end{bmatrix}$.

5. If A is an $n \times n$ matrix with $\det(A) = 2$: i) Find $\det(\text{adj}(A))$; ii) If $(\text{adj}(A))_{ij} = 1$, which entry of A^{-1} do you know the explicit value of?

6. Show that if A is symmetric, then $\text{adj}(A)$ is also symmetric.

7. Show that if $\det(A) = 0$, then $\det(\text{adj}(A)) = 0$. (Hint: Note that

$$\text{adj}(A)A = A\text{adj}(A) = \det(A)I,$$

and show that $\text{adj}(A)$ has a nontrivial null space if $\det(A) = 0$.)

8. Let $A = \begin{bmatrix} a & b \\ c & d \end{bmatrix}$. Show that the area of the parallelogram determined by the vectors 0, $(a\ b)$, (c, d) and $(a + c, b + d)$ is $|\det(A)|$.

9. Let

$$A = \begin{bmatrix} 1 & x_1 & x_1^2 & \cdots & x_1^{n-1} \\ 1 & x_2 & x_2^2 & \cdots & x_2^{n-1} \\ \vdots & & \vdots & & \vdots \\ 1 & x_n & x_n^2 & \cdots & x_n^{n-1} \end{bmatrix}.$$

Find $\det(A)$.

10. Let $n \geq 2$, $n \in \mathbb{N}$ and A_n be an $n \times n$ tridiagonal matrix given by

$$A_n = \begin{bmatrix} 3 & 2 & 0 & 0 & \cdots & 0 \\ 1 & 3 & 2 & 0 & \cdots & 0 \\ 0 & 1 & 3 & 2 & \cdots & 0 \\ \vdots & \vdots & \ddots & \ddots & \ddots & \vdots \\ 0 & 0 & \cdots & 1 & 3 & 2 \\ 0 & 0 & \cdots & 0 & 1 & 3 \end{bmatrix}.$$

Compute D_n where $D_n = \det(A_n)$.

11. Let $x = (x_1, x_2, \cdots, x_n) \in \mathbb{R}^n$ and $y = (1, 1, \cdots, 1) \in \mathbb{R}^n$. Let $A = I + yx^T$. Find $\det(A)$.

12. Let $x = (1, 1, \cdots, 1) \in \mathbb{R}^n$. Let $A = xx^T - I$. Find $\det(A)$ and A^{-1}.

13. Let $\det(A) = a$, $\det(B) = b$. Compute

$$\det \begin{bmatrix} A & C \\ 0 & B \end{bmatrix}.$$

14. Use Cramer's rule to solve the following systems, if applicable.

1)

$$\begin{cases} x + 2y + 3z = 1 \\ x + 3y = 1 \\ y + z = 1; \end{cases}$$

2)

$$\begin{cases} 2y + 3z = 1 \\ x + 3y = -1 \\ x + y + z = 1; \end{cases}$$

3)

$$\begin{cases} x + 2y + 3z = 1 \\ 2x + 3y + 4z = 1 \\ 5x + 3y + 4z = -1; \end{cases}$$

4)

$$\begin{cases} x - 2y + 3z = 1 \\ x + 3y + 4z = -1 \\ y + z = 1. \end{cases}$$

15. Let $A = (a_{ij})$ be an $n \times n$ matrix and M_{ij} be the submatrix of A after the i-th row and the j-th column of A are deleted. Then

$\det(A)$

$= a_{ij}(-1)^{1+j} \det M_{1j} + \cdots + a_{ji}(-1)^{j+i} \det M_{ji} + \cdots + a_{ni}(-1)^{n+i} \det M_{ni}.$

Chapter 6

Eigenvalues and eigenvectors

6.1 Introduction to eigenvectors and eigenvalues

There are many occasions that we model the input x_k at the discrete instance $k \in \mathbb{N}$ and the output y_k at instance k with a linear relation

$$y_k = Ax_k,$$

where A is an $n \times n$ matrix. Then the output at any instant can be predicted by $y_k = A^k x_1$, and sometimes we are even interested in the existence of the limit $\lim_{k \to +\infty} y_k$. It is usually time consuming to compute A^k when k and the dimension n of A arc large. However, if A is diagonal, then A^k can be easily computed. Indeed, we have

$$
A = \begin{bmatrix} \lambda_1 & & & \mathbf{0} \\ & \lambda_2 & & \\ & & \ddots & \\ \mathbf{0} & & & \lambda_n \end{bmatrix}
\quad \text{and} \quad
A^k = \begin{bmatrix} d_1^k & & & \mathbf{0} \\ & d_2^k & & \\ & & \ddots & \\ \mathbf{0} & & & d_n^k \end{bmatrix}.
$$

Certainly not every matrix is diagonal. The question is how to reduce a non-diagonal matrix into a diagonal one. Since we will compute successive multiplication of A, we assume there exists an invertible matrix P such that

$$A = P^{-1}DP,$$

where D is diagonal. Then we have

$$A^2 = (P^{-1}DP)(P^{-1}DP) = P^{-1}D^2P,\ A^k = P^{-1}D^kP.$$

This means that if we are able to find an invertible matrix P such that $A = P^{-1}DP$, we can easily compute A^k with $A^k = P^{-1}D^kP$. Now the question is transformed into finding a decomposition of A with $A = P^{-1}DP$, where D is diagonal. To be more specific, let $P = [p_1 : p_2 : \cdots : p_n]$, $D = \text{diag}\{\lambda_1, \lambda_2, \cdots, \lambda_n\}$. Then $A = P^{-1}DP$ is equivalent to

$$A[p_1 : p_2 : \cdots : p_n] = [\lambda_1 p_1, \lambda_2 p_2, \cdots, \lambda_n p_n].$$

Namely, to achieve $A = P^{-1}DP$, we need to find the scalar-vector pairs (λ_i, p_i), $i = 1, 2, \cdots, n$ such that

$$Ap_i = \lambda_i p_i,$$

and p_i, $i = 1, 2, \cdots, n$ are linearly independent.

Definition 6.1.1. Let A be an $n \times n$ matrix. If there exists a **nonzero** $x \in \mathbb{C}^n$ such that
$$Ax = \lambda x$$
for some $\lambda \in \mathbb{C}$, we call $x \neq 0$ an **eigenvector** of A and λ the **eigenvalue** of A corresponding to x. If there are n linearly independent eigenvectors $p_1, p_2, \cdots, p_n$ of A with corresponding eigenvalues $\lambda_1, \lambda_2, \cdots, \lambda_n$, then the matrix $P = [p_1 : p_2 : \cdots : p_n]$ is such that

$$P^{-1}AP = \begin{bmatrix} \lambda_1 & & & \mathbf{0} \\ & \lambda_2 & & \\ & & \ddots & \\ \mathbf{0} & & & \lambda_n \end{bmatrix},$$

and we say that A is **diagonalizable.**

Now we develop methods for computing eigenvalues of A. Notice from the definition of eigenvectors that every eigenvector x is nonzero and satisfies

$$Ax = \lambda x \Leftrightarrow (A - \lambda I)x = 0,$$

which implies that $A - \lambda I$ is not invertible. Then it is necessary that every eigenvalue λ of A satisfies

$$\det(A - \lambda I) = 0,$$

which is a n-th degree polynomial of λ and which has n roots in the complex domain $\mathbb{C}$. Conversely, if $\lambda \in \mathbb{C}$ satisfies $\det(A - \lambda I) = 0$, then $A - \lambda I$ is not invertible and there exists nonzero x which satisfies

$$(A - \lambda I)x = 0 \Leftrightarrow Ax = \lambda x.$$

> **Theorem 6.1.2.** Let A be an $n \times n$ matrix. $\lambda \in \mathbb{C}$ is an eigenvalue of A if and only if
> $$\det(A - \lambda I) = 0.$$

For every $n \times n$ matrix A, we call $\det(A - \lambda I)$ the **characteristic polynomial** of A, and call $N(A - \lambda I)$ the **eigenspace** of A corresponding to the eigenvalue λ.

Example 6.1.3. Let A be the following matrix:
$$A = \begin{bmatrix} -1 & 0 & 1 \\ -6 & 2 & 3 \\ 0 & 0 & 1 \end{bmatrix}.$$

i) Find the eigenvalues λ_1, λ_2, λ_3 of A with $\lambda_1 \leq \lambda_2 \leq \lambda_3$;

ii) Find an eigenvector associated to each eigenvalue obtained in i) with 1 as its first non-zero component;

iii) Determine whether A is diagonalizable or not. If yes, find a matrix P such that $P^{-1}AP$ is diagonal.

Solution: i) We solve the characteristic equation $\det(\lambda I - A) = 0$ for the eigenvalues. That is

$$\det \begin{bmatrix} \lambda + 1 & 0 & 1 \\ 6 & \lambda - 2 & -3 \\ 0 & 0 & \lambda - 1 \end{bmatrix} = 0 \Rightarrow (\lambda + 1)(\lambda - 1)(\lambda - 2) = 0$$

$$\Rightarrow \lambda_1 = -1, \ \lambda_2 = 1, \ \lambda_3 = 2.$$

ii) We solve the homogeneous system $(\lambda I - A)x = 0$ with $x = (x_1, x_2, x_3) \in \mathbb{R}^3$ for an eigenvector corresponding to every $\lambda \in \{\lambda_1, \lambda_2, \lambda_3\}$.

• For $\lambda = \lambda_1 = -1$, we have

$$\begin{bmatrix} 0 & 0 & -1 \\ 6 & -3 & -3 \\ 0 & 0 & -2 \end{bmatrix} \begin{bmatrix} x_1 \\ x_2 \\ x_3 \end{bmatrix} = \begin{bmatrix} 0 \\ 0 \\ 0 \end{bmatrix} \Rightarrow \begin{bmatrix} 1 & -\frac{1}{2} & 0 \\ 0 & 0 & 1 \\ 0 & 0 & 0 \end{bmatrix} \begin{bmatrix} x_1 \\ x_2 \\ x_3 \end{bmatrix} = \begin{bmatrix} 0 \\ 0 \\ 0 \end{bmatrix}$$

$$\Rightarrow x_1 + \left(-\frac{1}{2}\right) x_2 = 0, \ x_3 = 0$$

$$\Rightarrow x = \begin{bmatrix} x_1 \\ x_2 \\ x_3 \end{bmatrix} = \begin{bmatrix} \frac{1}{2}t \\ t \\ 0 \end{bmatrix}, \ t \in \mathbb{R}.$$

Putting $t = 2$ in x, we obtain an eigenvector associated with the eigenvalue λ_1:

$$p_1 = \begin{bmatrix} 1 \\ 2 \\ 0 \end{bmatrix}.$$

- For $\lambda = \lambda_2 = 1$, we have

$$\begin{bmatrix} 2 & 0 & -1 \\ 6 & -1 & -3 \\ 0 & 0 & 0 \end{bmatrix} \begin{bmatrix} x_1 \\ x_2 \\ x_3 \end{bmatrix} = \begin{bmatrix} 0 \\ 0 \\ 0 \end{bmatrix} \Rightarrow \begin{bmatrix} 1 & 0 & -\frac{1}{2} \\ 0 & -1 & 0 \\ 0 & 0 & 0 \end{bmatrix} \begin{bmatrix} x_1 \\ x_2 \\ x_3 \end{bmatrix} = \begin{bmatrix} 0 \\ 0 \\ 0 \end{bmatrix}$$

$$\Rightarrow x_1 + \left(-\frac{1}{2}\right) x_3 = 0, \ x_2 = 0$$

$$\Rightarrow x = \begin{bmatrix} x_1 \\ x_2 \\ x_3 \end{bmatrix} = \begin{bmatrix} \frac{1}{2}t \\ 0 \\ t \end{bmatrix}, \ t \in \mathbb{R}.$$

Putting $t = 2$ in x, we obtain an eigenvector associated with the eigenvalue λ_2:

$$p_2 = \begin{bmatrix} 1 \\ 0 \\ 2 \end{bmatrix}.$$

- For $\lambda = \lambda_3 = 2$, we have

$$\begin{bmatrix} 3 & 0 & -1 \\ 6 & 0 & -3 \\ 0 & 0 & 1 \end{bmatrix} \begin{bmatrix} x_1 \\ x_2 \\ x_3 \end{bmatrix} = \begin{bmatrix} 0 \\ 0 \\ 0 \end{bmatrix} \Rightarrow \begin{bmatrix} 1 & 0 & 0 \\ 0 & 0 & 1 \\ 0 & 0 & 0 \end{bmatrix} \begin{bmatrix} x_1 \\ x_2 \\ x_3 \end{bmatrix} = \begin{bmatrix} 0 \\ 0 \\ 0 \end{bmatrix}$$

$$\Rightarrow x_1 = 0, \ x_3 = 0$$

$$\Rightarrow x = \begin{bmatrix} x_1 \\ x_2 \\ x_3 \end{bmatrix} = \begin{bmatrix} 0 \\ t \\ 0 \end{bmatrix}, \ t \in \mathbb{R}.$$

Putting $t = 1$ in x, we obtain an eigenvector associated with the eigenvalue λ_3:

$$p_3 = \begin{bmatrix} 0 \\ 1 \\ 0 \end{bmatrix}.$$

iii) Let $P = [p_1 : p_2 : p_3]$. Then $\det(P) = -2 \neq 0$. Then A has three linearly independent eigenvectors p_1, p_2, p_3 and is diagonalizable. We have

$$P = \begin{bmatrix} 1 & 1 & 0 \\ 2 & 0 & 1 \\ 0 & 2 & 0 \end{bmatrix}$$

which satisfies that

$$P^{-1}AP = \begin{bmatrix} -1 & 0 & 0 \\ 0 & 1 & 0 \\ 0 & 0 & 2 \end{bmatrix}.$$

$\square$

Notice that $\det(\lambda I - A)|_{\lambda=0} = (-1)^n \det(A)$. Then by Theorem 6.1.2, we have

Theorem 6.1.4. Let A be an $n \times n$ matrix. A is invertible if and only if $\lambda = 0$ is not an eigenvalue of A.

Example 6.1.5.

A projection matrix P with $P^2 = P, P^T = P$ is not invertible and $\lambda = 0$ is always an eigenvalue.

An orthogonal matrix Q with $Q^T Q = I$ is invertible and $\lambda = 0$ is not an eigenvalue.

Exercise 6.1.6.

1. For the following matrices, find the eigenvalues and a basis of the associated eigenspaces.

$$A = \begin{bmatrix} 1 & 0 & 0 \\ 0 & 1 & 0 \\ 0 & 0 & 1 \end{bmatrix}, B = \begin{bmatrix} 1 & 1 & 1 \\ 1 & 1 & 1 \\ 1 & 1 & 1 \end{bmatrix}, C = \begin{bmatrix} 1 & 1 & 0 \\ 0 & 1 & 1 \\ 0 & 0 & 1 \end{bmatrix}.$$

2. Find a 3×3 matrix A whose eigenvalues are $-1, 2, 3$ and the corresponding eigenvectors are columns of

$$P = [c_1 : c_2 : c_3] = \begin{bmatrix} 1 & 1 & 0 \\ 0 & 1 & 1 \\ 0 & 0 & 1 \end{bmatrix},$$

respectively.

3. Find a 3×3 matrix A whose eigenvalues are $-1, 2, 2$ and the corresponding eigenvectors are columns of

$$P = [c_1 : c_2 : c_3] = \begin{bmatrix} 1 & 1 & 0 \\ 0 & 1 & 1 \\ 0 & 0 & 1 \end{bmatrix},$$

respectively.

4. Let A be an $n \times n$ matrix. Show that if λ is an eigenvalue of A and $\mathbf{x}$ a corresponding eigenvector, then for every $k \in \mathbb{N}$, λ^k is an eigenvalue of A^K with $\mathbf{x}$ a corresponding eigenvector.

5. Let A be an $n \times n$ invertible matrix. Show that if λ is an eigenvalue of A and $\mathbf{x}$ a corresponding eigenvector, then for every $k \in \mathbb{Z}$, λ^k is an eigenvalue of A^K with $\mathbf{x}$ a corresponding eigenvector.

6. Let f be a polynomial and A an $n \times n$ matrix. Show that if λ is an eigenvalue of A and $\mathbf{x}$ a corresponding eigenvector, then $f(\lambda)$ is an eigenvalue of $f(A)$ with $\mathbf{x}$ a corresponding eigenvector.

7. Let f and g be polynomials and A an $n \times n$ matrix. Show that

$$(f + g)(A) = f(A) + g(A),$$
$$(fg)(A) = f(A)g(A).$$

8. Let A be an $n \times n$ matrix. Show that there exists a nonzero polynomial f such that $f(A) = 0$.

9. Let A and P be $n \times n$ matrices and P is invertible. Show that for every polynomial f,
$$f(P^{-1}AP) = P^{-1}f(A)P.$$

10. Let A and P be $n \times n$ real matrices and P is invertible. Show that there exists a $\lambda \in \mathbb{C}$ such that $A + \lambda P$ is not invertible.

11. Let A and P be $n \times n$ real matrices and P is invertible. Show that if $n \in \mathbb{N}$ is odd, then there exists a $\lambda \in \mathbb{R}$ such that $A + \lambda P$ is not invertible.

12. Let A be an $n \times n$ diagonalizable matrix. Show that if every eigenvalue of A satisfies $|\lambda| < 1$, then
$$\lim_{k \to \infty} \det(A^k) = 0.$$

6.2 Diagonalizability

We have seen in Section 6.1 that if an $n \times n$ matrix A is diagonalizable, namely, A has n linearly independent eigenvectors, then the powers of A can be easily obtained. However, not every matrix A has n linearly independent eigenvectors. For example the matrix $J = \begin{bmatrix} 1 & 1 \\ 0 & 1 \end{bmatrix}$ has an eigenvalue $\lambda = 1$ with algebraic multiplicity 2 as a root of the characteristic polynomial $p(\lambda) = (\lambda - 1)^2$. However the null space of $A - 1 \cdot I$ is

$$N(A - I) = \text{span}\left(\begin{bmatrix} 1 \\ 0 \end{bmatrix} \right),$$

which is one dimensional. That is, we cannot find two linearly independent eigenvectors necessary to diagonalize A. Therefore, A is non-diagonalizable.

Definition 6.2.1. Let A be an $n \times n$ matrix. $\lambda_0 \in \mathbb{C}$ is an eigenvalue. We call the order of the factor $\lambda - \lambda_0$ in the characteristic polynomial $\det(A - \lambda_0)$ the **algebraic multiplicity** of λ_0. We call the dimension of the nullspace $N(A - \lambda_0 I)$ the **geometrical multiplicity** of λ_0.

Example 6.2.2. For the matrix $J = \begin{bmatrix} 1 & 1 \\ 0 & 1 \end{bmatrix}$, the algebraic multiplicity of the eigenvalue $\lambda_0 = 1$ is 2 because the order of the factor $\lambda - 1$ in the characteristic polynomial $\det(A - \lambda_0) = (\lambda - 1)^2$ is 2. The geometrical multiplicity of the eigenvalue $\lambda_0 = 1$ is 1 because

$$\dim N(A - I) = \dim \operatorname{span}\left(\begin{bmatrix} 1 \\ 0 \end{bmatrix}\right) = 1.$$

$\square$

Theorem 6.2.3. Let A be an $n \times n$ matrix. Let $\lambda_1, \lambda_2, \cdots, \lambda_k$ be a set of distinct eigenvalues of A. Let $S_1 = \{v_{11}, v_{12}, \cdots, v_{1n_1}\}, S_2 = \{v_{21}, v_{22}, \cdots, v_{2n_2}\}, \cdots, S_k = \{v_{k1}, v_{k2}, \cdots, v_{kn_k}\}$ be bases of the nullspaces $N(A - \lambda_j I)$, $j = 1, 2, \cdots, k$. Then the union of the bases $S_1 \cup S_2 \cup \cdots S_k$ is linearly independent.

Proof. We first show that $S_1 \cup S_2$ is linearly independent. Regard S_j as a matrix of basis vectors contained in S_j. Let $c_j = (c_{11}, c_{12}, \cdots, c_{1n_j})$, $j = 1, 2, \cdots, k$ and consider

$$S_1 c_1 + S_2 c_2 = 0. \tag{6.1}$$

Suppose, for the sake of contradiction, that $S_1 \cup S_2$ is linearly dependent, then $c_1 \neq 0$ and $c_2 \neq 0$; otherwise, one of S_1 and S_2 is linearly dependent which is impossible. Multiplying both sides of (6.1) by A, we obtain that

$$\lambda_1 S_1 c_1 + \lambda_2 S_2 c_2 = 0.$$

Multiplying both sides of $S_1 c_1 + S_2 c_2 = 0$ by λ_1 and subtracting it from the above equality we have

$$(\lambda_2 - \lambda_1) S_2 c_2 = 0,$$

which leads to $c_2 = 0$. This is a contradiction. Therefore, $S_1 \cup S_2$ is linearly independent.

Suppose that $S_1 \cup S_2 \cup \cdots S_j$, $j < k$ is linearly independent. Consider

$$S_1 c_1 + S_2 c_2 + \cdots + S_j c_j + S_{j+1} c_{j+1} = 0. \tag{6.2}$$

Multiplying both sides of (6.2) by A, we obtain that

$$\lambda_1 S_1 c_1 + \lambda_2 S_2 c_2 + \cdots + \lambda_j S_j c_j + \lambda_{j+1} S_{j+1} c_{j+1} = 0. \tag{6.3}$$

Multiplying both sides of (6.2) by λ_{j+1} and subtracting it from (6.3) we have

$$(\lambda_1 - \lambda_{j+1})S_1c_1 + (\lambda_2 - \lambda_{j+1})S_2c_2 + \cdots + (\lambda_j - \lambda_{j+1})S_jc_j = 0. \qquad (6.4)$$

Then $c_1 = c_2 = \cdots = c_j = 0$ since $S_1 \cup S_2 \cup \cdots S_j$, $j < k$ is linearly independent. Then by (6.2) we have $c_{j+1} = 0$ and hence $S_1 \cup S_2 \cup \cdots S_j \cup S_{j+1}$ is linearly independent.

By mathematical induction, $S_1 \cup S_2 \cup \cdots \cup S_k$ is linearly independent. $\qquad \square$

An immediate consequence of Theorem 6.2.3 is that

Corollary 6.2.4. Let A be an $n \times n$ matrix.

i) If A has n distinct eigenvalues, then A is diagonalizable.

ii) Eigenvectors associated with distinct eigenvalues are linearly independent.

A necessary and sufficient condition for diagonalizability is the following

---

Theorem 6.2.5. Let A be an $n \times n$ matrix. Then

i) for every eigenvalue λ_0 of A,

$$\text{algebraic multiplicity of } \lambda_0 \geq \text{geometrical multiplicity of } \lambda_0.$$

ii) A is diagonalizable if and only if for every eigenvalue λ_0,

$$\text{algebraic multiplicity of } \lambda_0 = \text{geometrical multiplicity of } \lambda_0.$$

Proof. i) Let m be the geometrical multiplicity of λ_0. Let $S_{\lambda_0} = \{v_1, v_2, \cdots, v_m\}$ be a basis of $N(A - \lambda_0 I)$. Then we can extend S_{λ_0} into a basis $\{v_1 : v_2 : \cdots : v_m : v_{m+1} : \cdots : v_n\}$ of $\mathbb{R}^n$. Then we have

$$A[v_1 : v_2 : \cdots : v_m : v_{m+1} : \cdots : v_n]$$

$$= [v_1 : v_2 : \cdots : v_m : v_{m+1} : \cdots : v_n] \begin{bmatrix} \lambda_0 I_{m \times m} & B \\ 0 & C \end{bmatrix},$$

where I_m is the $m \times m$ identity matrix, B is an $m \times (n-m)$ matrix and C is an $(n-m) \times (n-m)$ matrix.

Let $P = [v_1 : v_2 : \cdots : v_m : v_{m+1} : \cdots : v_n]$. Then we have

$$P^{-1}AP = \begin{bmatrix} \lambda_0 I_m & B \\ 0 & C \end{bmatrix},$$

and

$$P^{-1}(A - \lambda I)P = \begin{bmatrix} (\lambda_0 - \lambda)I_m & B \\ 0 & C - \lambda I_{(n-m)} \end{bmatrix}.$$

Then we have

$$\det(A - \lambda I) = \det((\lambda_0 - \lambda)I_m) \det(C - \lambda I_{(n-m)})$$
$$= (\lambda - \lambda_0)^m \det(C - \lambda I_{(n-m)}).$$

Therefore, we obtain that

algebraic multiplicity of $\lambda_0 \geq m =$ geometrical multiplicity of λ_0.

ii) "$\Longrightarrow$" If A is diagonalizable, then A has n linearly independent eigenvectors. Note that the sum of algebraic multiplicities of all eigenvalues is n. If there exists an eigenvalue λ_0 with algebraic multiplicity of λ_0 strictly larger than the geometrical multiplicity of λ_0, then the sum of the dimensions of all the eigenspaces is less than n. By Theorem 6.2.3, A has less than n linearly independent eigenvectors. This is a contradiction.

"$\Longleftarrow$" If for every eigenvalue λ_0,

algebraic multiplicity of λ_0 = geometrical multiplicity of λ_0

then the sum of the dimensions of all the eigenspaces is n. By Theorem 6.2.3, A has n linearly independent eigenvectors. A is diagonalizable. $\square$

Similar matrices

The idea of diagonalizing a square matrix extends to the situation that a matrix is not diagonalizable. For a given matrix A, we say that B is **similar** to A if there exists an invertible matrix P such that $B = P^{-1}AP$. It is evident that B is similar to A is equivalent to that A is similar to B. If A is similar to B, we write $A \sim B$. It turns out similar matrices share many important properties which are listed below:

> **Theorem 6.2.6.** Let A, B be $n \times n$ matrices with $A = P^{-1}BP$. Then we have
>
> i) $\det(A) = \det(B)$ and $\det(A - \lambda I) = \det(B - \lambda I)$;
>
> ii) $\text{rank}(A) = \text{rank}(B)$;
>
> iii) $\text{trace}(A) = \text{trace}(B)$, where $\text{trace}(A) = \sum_{i=1}^{n} a_{ii}$.

Proof. i) We have

$$\det(A) = \det(P^{-1}BP)$$
$$= \det(P^{-1}) \det(B) \det(P)$$
$$= \frac{1}{\det(P)} \det(B) \det(P)$$

$$= \det(B).$$

Moreover, we have

$$\begin{aligned}
\det(A - \lambda I) &= \det(P^{-1}BP - \lambda I) \\
&= \det(P^{-1}(B - \lambda I)P) \\
&= \det(B - \lambda I).
\end{aligned}$$

ii) By Theorem 3.4.6, we have

$$\mathrm{rank}(A) \le \min\{\mathrm{rank}(P^{-1}), \ \mathrm{rank}(B), \ \mathrm{rank}(P)\} \le \mathrm{rank}(B).$$

Since $B = PAP^{-1}$, we also have

$$\mathrm{rank}(B) \le \min\{\mathrm{rank}(P^{-1}), \ \mathrm{rank}(A), \ \mathrm{rank}(P)\} \le \mathrm{rank}(A).$$

Therefore, we have $\mathrm{rank}(A) = \mathrm{rank}(B)$.

iii) We first show that for every $m \times n$ matrix C and $n \times m$ matrix D, we have

$$\begin{aligned}
\mathrm{trace}(CD) &= \sum_{i=1}^{m}(CD)_{ii} \\
&= \sum_{i=1}^{m}(\text{row } i \text{ of } C) \cdot (\text{column } i \text{ of } D) \\
&= \sum_{i=1}^{m}\sum_{j=1}^{n} C_{ij}D_{ji} \\
&= \sum_{j=1}^{n}\sum_{i=1}^{m} D_{ji}C_{ij} \\
&= \sum_{j=1}^{n}(DC)_{jj} \\
&= \mathrm{trace}(DC).
\end{aligned}$$

Therefore, we have $\mathrm{trace}(A) = \mathrm{trace}(P^{-1}BP) = \mathrm{trace}(PP^{-1}B) = \mathrm{trace}(B)$. $\qquad\square$

Example 6.2.7. Let $A = \begin{bmatrix} 1 & 2 \\ 3 & 4 \end{bmatrix}$, $P = \begin{bmatrix} 0 & 1 \\ 1 & 0 \end{bmatrix}$, $B = P^{-1}AP = \begin{bmatrix} 4 & 3 \\ 2 & 1 \end{bmatrix}$. Then we have i) $\det(A) = -2 = \det(B)$, $\det(A - \lambda I) = \lambda^2 - 5\lambda - 2 = \det(B - \lambda I)$; ii) $\mathrm{rank}(A) = \mathrm{rank}(B) = 2$; iii) $\mathrm{trace}(A) = 5 = \mathrm{trace}(B)$. $\qquad\square$

Estimates of eigenvalues

Usually elementary row operations do not preserve eigenvalues. For example the permutation matrix $E = \begin{bmatrix} 0 & 1 \\ 1 & 0 \end{bmatrix}$ has eigenvalues $\lambda = \pm 1$. But $EE = I$ has eigenvalues $\lambda_{12} = 1$ only. However, we still have certain estimates for the information of the eigenvalues without practically computing them. Consider the characteristic polynomial

$$\det(A - \lambda I),$$

and assume it can be factored into

$$\det(A - \lambda I) = (\lambda_1 - \lambda)(\lambda_2 - \lambda) \cdots (\lambda_n - \lambda),$$

where λ_i, $i = 1, 2, \cdots, n$ are eigenvalues of A. Putting $\lambda = 0$, then we have

$$\det(A) = \lambda_1 \lambda_2 \cdots \lambda_n.$$

The next estimate on eigenvalues makes use of the fact that every square matrix A can be triangularized by an invertible matrix P such that

$$P^{-1}AP = D,$$

where D is either a diagonal matrix with eigenvalues of A in the main diagonal, or D is the **Jordan form** (see Theorem 6.3.8) of A which is a triangular matrix with eigenvalues of A in the main diagonal. Note repeated eigenvalues all appear in the main diagonal of D. Then we have

$$\text{trace}(D) = \text{trace}(P^{-1}AP) = \text{trace}(PP^{-1}A) = \text{trace}(A).$$

In summary, we arrived at

$$\text{trace}(A) = \lambda_1 + \lambda_2 + \cdots + \lambda_n,$$

where λ_i, $i = 1, 2, \cdots, n$ are eigenvalues of A.

Another estimate of eigenvalues makes use of the dominant matrix. Recall that every dominant matrix is invertible (See Example 2.3.8). However, we know that if λ_0 is an eigenvalue of A, $A - \lambda_0 I$ is not invertible and hence not dominant. That is, there exists a row number $k \in \{1, 2, \cdots, n\}$ such that λ_0 is in the following so-called **Gershgorin disc**:

$$|a_{kk} - \lambda_0| \leq \sum_{i=1, \, i \neq k}^{n} |a_{ki}|.$$

That is,

Theorem 6.2.8 (Gershgorin's disc theorem). Let A be an $n \times n$ matrix. Then every eigenvalue λ_0 of A lies in at least one of the circles around the main diagonal entry:

$$\lambda_0 \in \cup_{i=1}^n D_i,$$

where

$$D_i = \{z \in \mathbb{C} : |z - a_{kk}| \leq \sum_{i=1, \, i \neq k}^n |a_{ki}|, \, i \in \{1, 2, \cdots, n\}.$$

Note that since A and A^T share the same set of eigenvalues, Gershgorin's disc theorem also applies with Gershgorin's discs obtained according to the columns of A.

Example 6.2.9. Consider $A = \begin{bmatrix} 1 & -1 & 0 \\ -1 & 2 & 0 \\ 2 & -1 & 4 \end{bmatrix}$. Then we have $D_1 = \{z \in \mathbb{C} : |z - 1| \leq 1\}, D_2 = \{z \in \mathbb{C} : |z - 2| \leq 1\}$ and $D_3 = \{z \in \mathbb{C} : |z - 4| \leq 3\}$. Then every eigenvalue of A is in the union $D_1 \cup D_2 \cup D_3$.

Theorem 6.2.10 (Gershgorin's second disc theorem). **(optional)** Let A be an $n \times n$ matrix. A subset S of the Gershgorin discs is called a disjoint group of discs if no disc in the group S intersects a disc which is not in S. If a disjoint group S contains r nonconcentric discs, then there are r eigenvalues in S.

Proof. Let $A(t)$ be the matrix obtained from A with the off diagonal elements multiplied by the variable t, where $t \in [0, 1]$. Note $A(0)$ is the diagonal matrix with the same diagonal of A, and $A(1) = A$.

Then $A(0)$ has the n main diagonal entries of the eigenvalues and the n Gershgorin discs are themselves. As t ranges from 0 to 1, the Gershgorin discs will also change its radius with centers fixed at the main diagonals. Moreover, the eigenvalues will also change inside the Gershgorin discs.

Since the roots of the characteristic polynomial $\det(A(t) - \lambda I)$ of $A(t)$ change continuously with respect to $t \in [0, 1]$, the traces of the eigenvalues are continuous curves inside the Gershgorin discs. If a disjoint group S contains r nonconcentric discs, then there are r eigenvalues which never escape from the group S during the change of t from 0 to 1. $\square$

Cayley–Hamilton theorem (optional)

Let $p(x) = a_0 + a_1 x + a_2 x^2 + \cdots + a_m x^m$ be a polynomial and A an $n \times n$ matrix. If A has an eigenvalue λ and eigenvector $u \in \mathbb{C}^n$, how to obtain eigenvalue and eigenvector of $p(A) = a_0 I + a_1 A + a_2 A^2 + \cdots + a_m A^m$? Note that we have

$$Au = \lambda u$$
$$A^2 u = \lambda^2 u$$
$$\cdots$$
$$A^m u = \lambda^m u.$$

Then we have

$$p(A)u = p(\lambda)u.$$

Namely, every eigenvector of A is an eigenvector of the polynomial $p(A)$ of A, and if λ is an eigenvalue of A then $p(\lambda)$ is an eigenvalue of $p(A)$. It is then interesting to ask whether it is possible that for some polynomial p, $p(A)$ has eigenvectors which are not that of A. It turns out the **Cayley-Hamilton theorem** claims that there exists a polynomial p such that $p(A) = 0$, which means every nonzero vector x is an eigenvector of $p(A)$. Therefore, there exists eigenvectors of $p(A)$ which are not that of A.

Suppose that p is the characteristic polynomial of A, and A is diagonalizable. Then we have n linearly independent eigenvectors and $p(\lambda) = 0$. It follows that $p(A) = 0$ because the eigenvectors form a basis for $\mathbb{R}^n$. Moreover, it seems that $p(A) = 0$ holds true even if A is not diagonalizable. For example, $p(\lambda) = (\lambda - 1)^2$ is the characteristic polynomial of

$$J = \begin{bmatrix} 1 & 1 \\ 0 & 1 \end{bmatrix},$$

and J is not diagonalizable. But $p(J) = (J - I)^2 = 0$.

Consider the adjoint of $\lambda I - A$. Let $B = \mathrm{adj}(\lambda I - A)$. By definition of adjoint, B can be represented as

$$B = \sum_{i=0}^{n-1} \lambda^i B_i.$$

We have

$$(\lambda I - A)\mathrm{adj}(\lambda I - A) = \det(\lambda I - A)I = p(\lambda)I,$$

where $p(\lambda) = a_0 + a_1 \lambda + a_2 \lambda^2 + \cdots + a_n \lambda^n$ is the characteristic polynomial of A. Then we have

$$\begin{aligned} p(\lambda)I &= (a_0 + a_1 \lambda + a_2 \lambda^2 + \cdots + a_n \lambda^n)I \\ &= (\lambda I - A)\mathrm{adj}(\lambda I - A) \\ &= (\lambda I - A)B \\ &= (\lambda I - A)\sum_{i=0}^{n-1} \lambda^i B_i \end{aligned}$$

$$= \lambda^n B_{n-1} + \sum_{i=1}^{n-1} \lambda^i (B_{i-1} - AB_i) - AB_0.$$

By comparing coefficient matrices of λ^i, $i = 0, 1, 2, \cdots, n$, we have

$$B_{n-1} = a_n I,$$
$$B_{i-1} - AB_i = a_i I, \ 1 \le i \le n - 1,$$
$$-AB_0 = a_0 I.$$

Then multiplying both sides of the above equalities by A^n, A^{n-1},...A^0, respectively, we have a telescoping sum on the left hand side that

$$A^n B_{n-1} + A^{n-1}(B_{n-2} - AB_{n-1}) + A^{n-2}(B_{n-2} - AB_{n-2}) + \cdots$$
$$+ A(B_0 - AB_1) - AB_0 = 0,$$

but the right hand side is $p(A)$. Therefore, we have $p(A) = 0$.

> **Theorem 6.2.11.** Let A be an $n \times n$ matrix with characteristic polynomial $p(\lambda)$. Then we have
>
> $$p(A) = 0.$$

The Cayley-Hamilton theorem has many applications. The following examples shed some light on how it can be applied.

Example 6.2.12. Let $A = \begin{bmatrix} 1 & 3 & 5 \\ 0 & 1 & 0 \\ 1 & 0 & 2 \end{bmatrix}$. The characteristic polynomial is $p(\lambda) = \lambda^3 - 4\lambda^2 + 5\lambda + 3$. By the Cayley-Hamilton theorem we have

$$A^3 - 4A^2 + 5A + 3I = 0.$$

which gives an expression of the highest term A^3 with

$$A^3 = 4A^2 - 5A - 3I,$$

and an expression for the inverse of A,

$$A^{-1} = -\frac{1}{3}\left(A^2 - 4A + 5I\right).$$

$\square$

Exercise 6.2.13.

1. Determine the diagonalizability of the following matrices:

$$A = \begin{bmatrix} 1 & 1 & 0 \\ 0 & 1 & 1 \\ 0 & 0 & 0 \end{bmatrix}, \ B = \begin{bmatrix} 1 & 1 & 0 \\ 0 & 1 & 0 \\ 0 & 0 & 1 \end{bmatrix}, \ C = \begin{bmatrix} 1 & 1 & 0 \\ 0 & 1 & 1 \\ 0 & 0 & 1 \end{bmatrix}.$$

2. Let $P^2 = P$. Is it true that $\lambda = 1$ and $\lambda = 0$ are both eigenvalues of P? Justify your answer.

3. Let A, B be $n \times n$ matrices. Show that AB and BA have the same characteristic polynomial. Hint: Use block eliminations on the matrix $\begin{bmatrix} A & \lambda I \\ I & B \end{bmatrix}$ to produce $\lambda I - AB$ and $\lambda I - BA$, respectively.

4. Let A, B be similar $n \times n$ matrices with $A = P^{-1}BP$. Show that if λ_0 is an eigenvalue of A with $\mathbf{x}_0$ a corresponding eigenvector, then λ_0 is an eigenvalue of B with $P\mathbf{x}_0$ a corresponding eigenvector.

5. Let $A = \begin{bmatrix} 2 & 2 & 2 \\ 2 & 2 & 2 \\ 2 & 2 & 2 \end{bmatrix}$ and $B = \begin{bmatrix} 0 & 0 & 0 \\ 0 & 0 & 0 \\ 0 & 0 & 6 \end{bmatrix}$. Show that A and B are similar.

6. Let $A = \begin{bmatrix} 1 & 2 & 3 \\ 0 & 1 & 2 \\ 3 & 1 & 0 \end{bmatrix}$ and $B = \begin{bmatrix} 1 & 2 & 0 \\ 1 & 0 & 3 \\ 2 & 3 & 1 \end{bmatrix}$. Show that A and B are similar.

7. Let A, B be similar $n \times n$ matrices. Show that

i) A and B have the same nullity, namely, $\dim N(A) = \dim N(B)$;

ii) A and B have the same set of eigenvalues;

iii) If λ_0 is an eigenvalue of A, then $\dim N(A - \lambda_0 I) = \dim N(B - \lambda_0 I)$;

iv) A is invertible if and only if B is invertible;

v) A is diagonalizable if and only if B is diagonalizable.

8. Let A be an $n \times n$ matrix. Show that A and A^T have the same set of eigenvalues.

9. Let A, B be $n \times n$ matrices. Show that AB and BA have the same set of eigenvalues.

10. Show that similarity is an equivalence relation on M_{nn}. Namely,

i) (Reflectivity) For every $A \in M_{nn}$, $A \sim A$;

ii) (Symmetry) For every A, $B \in M_{nn}$, $A \sim B$ implies $B \sim A$;

iii) (Transitivity) For every A, B, $C \in M_{nn}$, $A \sim B$ and $B \sim C$ imply $A \sim C$.

11. Find the Gershgorin's discs for possible location of each of the eigenvalues of each of the following matrices.

$$A = \begin{bmatrix} 1 & 0 & 0 \\ 0 & 4 & 0 \\ 0 & 0 & 6 \end{bmatrix}, \quad B = \begin{bmatrix} 1 & 2 & 3 \\ 3 & 4 & -5 \\ 4 & 5 & 0 \end{bmatrix}, \quad C = \begin{bmatrix} 1 & 2 & 0 \\ 3 & 0 & 2 \\ 0 & 3 & 6 \end{bmatrix}, \quad D = \begin{bmatrix} 1 & 2 & 0 \\ 0 & 4 & 2 \\ 0 & 0 & 6 \end{bmatrix}.$$

12. Let $A = \begin{bmatrix} 1 & 2 & 3 \\ 3 & 1 & 2 \\ 2 & 3 & 1 \end{bmatrix}$. Which ones of the following discs in $\mathbb{C}$ may contain eigenvalues of A?

a) $\{z \in \mathbb{C} : |z + 4| \leq 10\}$;

b) $\{z \in \mathbb{C} : |z - 6| \leq 10\}$;

c) $\{z \in \mathbb{C} : |z| \leq 6\}$;

d) $\{z \in \mathbb{C} : |z + 6| \leq 1\}$.

13. Let A and B be $n \times n$ real matrices, and A is diagonalizable such that $P^{-1}AP$ is diagonal for some invertible matrix P. Let r be the value $\|P^{-1}BP)\|_\infty$, where $\| \cdot \|_\infty$ is the **infinity norm** of $n \times n$ matrices defined by

$$\|A\|_\infty = \max_{1 \leq i \leq n} \sum_{j=1}^{n} |a_{ij}|.$$

Use Gershgorin's theorem to show that for every eigenvalue λ_{A+B} of $A + B$, there exists an eigenvalue λ_A of A, such that

$$|\lambda_{A+B} - \lambda_A| \leq r.$$

14. Let A be an $n \times n$ real matrix. Show that if $\|A\|_\infty < 1$, then $I - A$ is invertible and

$$(I - A)^{-1} = \sum_{n=1}^{\infty} A^n = A + A^2 + \cdots + A^n + \cdots.$$

15. Let A be an $m \times n$ matrix and $x \in \mathbb{R}^n$. Show that

$$\|Ax\|_\infty \leq \|A\|_\infty \|x\|_\infty.$$

16. Use the Cayley-Hamilton theorem to compute A^{-1} if

$$\begin{bmatrix} 1 & 1 & 0 \\ 0 & 1 & 1 \\ 0 & 0 & 1 \end{bmatrix}.$$

6.3 Applications to differential equations

Recall that the power series expansion of e^t converges uniformly on every closed interval $[-L, L]$, $L > 0$. Namely

$$e^t = \sum_{n=0}^{\infty} \frac{t^n}{n!} = 1 + \frac{t}{1!} + \frac{t^2}{2!} + \cdots + \frac{t^n}{n!} + \cdots,$$

for every $|t| \leq L$. If we have definition of metric (or magnitude) for matrices in M_{nn}, for instance, define the magnitude of A by $\|A\| = \left(\sum_{i,j}^{n} a_{ij}^2\right)^{\frac{1}{2}}$, we can define e^A, $A \in M_{nn}$ by the power series

$$e^A = \sum_{n=0}^{\infty} \frac{A^n}{n!} = I + \frac{A}{1!} + \frac{A^2}{2!} + \cdots + \frac{A^n}{n!} + \cdots,$$

which converges uniformly for every $A \in M_{nn}$ with $\|A\| \leq L$.

Proposition 6.3.1. If $AB \in M_{nn}$ are such that $AB = BA$, then $e^{A+B} = e^A e^B$.

Proof. If $AB = BA$, then by the binomial formula we have

$$(A + B)^n = \sum_{k=0}^{n} \frac{n!}{(n-k)!k!} A^{n-k} B^k.$$

Moreover, we have

$$\begin{aligned}
e^{A+B} &= \sum_{n=0}^{\infty} \frac{1}{n!} \sum_{k=0}^{n} \frac{n!}{(n-k)!k!} A^{n-k} B^k \\
&= \sum_{j=0}^{\infty} \sum_{k=0}^{\infty} \frac{1}{j!k!} A^j B^k \\
&= \sum_{j=0}^{\infty} \frac{1}{j!} A^j \sum_{k=0}^{\infty} \frac{1}{k!} B^k \\
&= e^A e^B.
\end{aligned}$$

$\square$

The following matrix-valued function is welldefined in $\mathbb{R}$:

$$\mathbb{R} \ni t \rightarrow e^{At} \in M_{nn},$$

with

$$e^{At} = \sum_{n=0}^{\infty} \frac{(At)^n}{n!} = I + \frac{A}{1!} t + \frac{A^2}{2!} t^2 + \cdots + \frac{A^n}{n!} t^n + \cdots.$$

We have

$$\frac{d}{dt} e^{At} = A + \frac{A^2}{1!} t + \frac{A^3}{2!} t^2 + \cdots + \frac{A^n}{(n-1)!} t^{n-1} + \cdots = A e^{At} = e^{At} A.$$

For every $\mathbf{c} \in \mathbb{R}^n$, let $x(t) = e^{At}\mathbf{c}$. We have

$$\frac{\mathrm{d}}{\mathrm{d}t}x(t) = \frac{\mathrm{d}}{\mathrm{d}t}e^{At}\mathbf{c} = Ae^{At}\mathbf{c} = Ax(t).$$

Namely, $x(t) = e^{At}\mathbf{c}$ is a solution of the differential equation

$$\frac{\mathrm{d}}{\mathrm{d}t}x(t) = Ax(t),$$

with initial data $x(0) = \mathbf{c}$. Conversely, if $\frac{\mathrm{d}}{\mathrm{d}t}x(t) = Ax(t)$, with $x(t) \in \mathbb{R}^n$ we multiply both sides by e^{-At} to obtain

$$e^{-At}\frac{\mathrm{d}}{\mathrm{d}t}x(t) = e^{-At}Ax(t),$$

which leads to

$$\frac{\mathrm{d}}{\mathrm{d}t}\left(e^{-At}x(t)\right) = 0.$$

Hence $e^{-At}x(t)$ is a constant vector $\mathbf{c}$ for all $t \in \mathbb{R}$ which satisfies

$$e^{-At}x(t)\big|_{t=0} = \mathbf{c}.$$

That is, $\mathbf{c} = x(0)$.

Theorem 6.3.2. The solution of the system of linear differential equations

$$\begin{cases} \dfrac{\mathrm{d}}{\mathrm{d}t}x(t) = Ax(t) \\[2mm] \qquad x(0) = \mathbf{c} \end{cases}$$

is $x(t) = e^{At}\mathbf{c}$.

Remark 6.3.3. We call $\varphi(t) = e^{At}$ the **standard fundamental solution matrix** of the system of linear differential equations $x'(t) = Ax(t)$, which satisfies $\varphi(0) = I$. Note that for every invertible matrix $P \in M_{nn}$, $e^{At}P$ satisfies that

$$\frac{\mathrm{d}}{\mathrm{d}t}(e^{At}P) = A(e^{At}P).$$

Namely each column of $e^{At}P$ is a solution of $x'(t) = Ax(t)$. Therefore, we call $e^{At}P$ a **fundamental solution matrix**.

Remark 6.3.4. By Proposition 6.3.1, we have

$$e^{A(t+s)} = e^{At}e^{As},$$

for all $s,\,t \in \mathbb{R}$ and hence

$$e^{-At}e^{At} = I, \text{ for all } t \in \mathbb{R}.$$

$\square$

Since e^{At} is defined by an infinite series, it is in general not efficient to use a series to compute the exact solutions of differential equations. We need to find alternative methods for computing e^{At}.

A is diagonalizable

If A is diagonalizable, there exists an invertible matrix

$$P = [x_1 : x_2 : \cdots : x_n] \in M_{nn}$$

such that

$$P^{-1}AP = \Lambda = \begin{bmatrix} \lambda_1 & & & \mathbf{0} \\ & \lambda_2 & & \\ & & \ddots & \\ \mathbf{0} & & & \lambda_n \end{bmatrix}.$$

We know that x_i, $i = 1, 2, \cdots, n$ are eigenvectors of A associated with eigenvalues λ_i, $i = 1, 2, \cdots, n$, respectively. Then we have

$$\begin{aligned} e^{At} &= e^{P\Lambda P^{-1}t} \\ &= I + \frac{P\Lambda P^{-1}}{1!}t + \frac{P\Lambda^2 P^{-1}}{2!}t^2 + \cdots + \frac{P\Lambda^n P^{-1}}{n!}t^n + \cdots \\ &= Pe^{\Lambda t}P^{-1} \\ &= P \begin{bmatrix} e^{\lambda_1 t} & & & \mathbf{0} \\ & e^{\lambda_2 t} & & \\ & & \ddots & \\ \mathbf{0} & & & e^{\lambda_n t} \end{bmatrix} P^{-1}. \end{aligned}$$

Then the solution of $\frac{d}{dt}x(t) = Ax(t)$ with initial value $x(0) = \mathbf{c}$ can be written

$$\begin{aligned} x(t) &= Pe^{\Lambda t}P^{-1}\mathbf{c} \\ &= x_1 e^{\lambda_1 t}c_1 + x_2 e^{\lambda_2 t}c_2 + \cdots + x_n e^{\lambda_n t}c_n, \end{aligned}$$

where $\begin{bmatrix} c_1 \\ c_2 \\ \vdots \\ c_n \end{bmatrix} = P^{-1}\mathbf{c}$. In summary, we have

Theorem 6.3.5. If A is diagonalizable, there exists $P = [x_1 : x_2 : \cdots : x_n] \in M_{nn}$ such that

$$P^{-1}AP = \Lambda = \begin{bmatrix} \lambda_1 & & & \mathbf{0} \\ & \lambda_2 & & \\ & & \ddots & \\ \mathbf{0} & & & \lambda_n \end{bmatrix}.$$

The solution of the system of linear differential equations

$$\begin{cases} \dfrac{d}{dt}x(t) = Ax(t) \\ x(0) = \mathbf{c}, \end{cases}$$

is

$$x(t) = Pe^{\Lambda t}P^{-1}\mathbf{c} = x_1 e^{\lambda_1 t}c_1 + x_2 e^{\lambda_2 t}c_2 + \cdots + x_n e^{\lambda_n t}c_n,$$

where $\begin{bmatrix} c_1 & c_2 & \cdots & c_n \end{bmatrix}^T = P^{-1}\mathbf{c}$.

Example 6.3.6. Consider the system of linear equations

$$\begin{cases} \dot{x}_1 = x_1 + x_2, \\ \dot{x}_2 = 3x_1 - x_2. \end{cases}$$

We have

$$\frac{d}{dt}\begin{bmatrix} x_1 \\ x_2 \end{bmatrix} = \begin{bmatrix} 1 & 1 \\ 3 & -1 \end{bmatrix}\begin{bmatrix} x_1 \\ x_2 \end{bmatrix}.$$

The characteristic polynomial of $A = \begin{bmatrix} 1 & 1 \\ 3 & -1 \end{bmatrix}$ is $(\lambda - 1)(\lambda + 1) - 3$. The eigenvalues are $\lambda_1 = 2$, $\lambda_2 = -2$. Solving system $(A - \lambda_1)x = 0$ we obtain an eigenvector $x_1 = (1, 1)$. Similarly $(A - \lambda_2)x = 0$ leads to an eigenvector $x_2 = (1, -3)$. Therefore, the solution is

$$x(t) = x_1 e^{\lambda_1 t}c_1 + x_2 e^{\lambda_2 t}c_2 = \begin{bmatrix} 1 \\ 1 \end{bmatrix}e^{2t}c_1 + \begin{bmatrix} 1 \\ -3 \end{bmatrix}e^{-2t}c_2,$$

where (c_1, c_2) is determined by initial values of $x(t)$. For instance, if $x(0) = (1, 1)$, then we have

$$\begin{bmatrix} c_1 \\ c_2 \end{bmatrix} = [x_1 : x_2]^{-1}x(0) = \begin{bmatrix} 1 & 1 \\ 1 & -3 \end{bmatrix}^{-1}\begin{bmatrix} 1 \\ 1 \end{bmatrix} = \begin{bmatrix} 1 \\ 0 \end{bmatrix}.$$

Remark 6.3.7. If there is a complex eigenvalue $\lambda = a \pm bi$, we can replace the complex valued solutions $e^{(a \pm bi)t}$ with the following real-valued ones,

$$\frac{e^{at+bti} + e^{at-bti}}{2} = \frac{1}{2}e^{at}\cos bt, \text{ and } \frac{e^{at+bti} - e^{at-bti}}{2i} = \frac{1}{2}e^{at}\sin bt.$$

A is not diagonalizable (optional)

If A is not diagonalizable, then there exists at least one eigenvalue with geometrical multiplicity less than its algebraic multiplicity. Then A is similar to a matrix J which is called the **Jordan form**. (See Chapter 8 for details.)

> **Theorem 6.3.8.** If $A \in M_{nn}$ has $s < n$ linearly independent eigenvectors, it is similar to a matrix J which has s Jordan blocks $J_1, J_2, \cdots, J_s$, where
>
> $$J = \begin{bmatrix} J_1 & & & \\ & J_2 & & \\ & & \ddots & \\ & & & J_s \end{bmatrix}, \quad J_j = \begin{bmatrix} \lambda_i & 1 & & \\ & \lambda_i & 1 & \\ & & \ddots & 1 \\ & & & \lambda_i \end{bmatrix},$$
>
> $j = 1, 2, \cdots, s$, and the size n_j of J_j is the algebraic multiplicity of the eigenvalue λ_j with $n_1 + n_2 + \cdots + n_s = n$.

Let $A = PJP^{-1}$, where J is the Jordan form of A. We have

$$\begin{aligned} e^{At} &= Pe^{Jt}P^{-1} \\ &= P \begin{bmatrix} e^{J_1 t} & & & \\ & e^{J_2 t} & & \\ & & \ddots & \\ & & & e^{J_s t} \end{bmatrix} P^{-1}. \end{aligned}$$

Notice that for every Jordan block J_j it can be written as $J_j = \lambda_j I + (J_j - \lambda_j I)$ with

$$J_j = \begin{bmatrix} \lambda_i & & & \\ & \lambda_i & & \\ & & \ddots & \\ & & & \lambda_i \end{bmatrix} + \begin{bmatrix} 0 & 1 & & \\ & 0 & 1 & \\ & & \ddots & 1 \\ & & & 0 \end{bmatrix}.$$

Moreover, the line of 1's in $N_j = (J_j - \lambda_j I)$ moves parallel with the main diagonal to the upper right corner when the power of N_j^k increases from $k = 1, 2, \cdots, n_j$, and $N_j^{n_j} = 0$. Notice that identity matrix multiplication is commutative with any matrix. We have

$$e^{J_j t} = e^{\lambda_j t + N_j t} = e^{\lambda_j t}\left(I + tN_j + \frac{t^2}{2}N_j^2 + \cdots + \frac{t^{n_j - 1}}{(n_j - 1)!}N_j^{n_j - 1} \right)$$

$$=e^{\lambda_j t}\begin{bmatrix} 1 & t & \frac{t^2}{2!} & \cdots & \frac{t^{n_j-1}}{(n_j-1)!} \\ & 1 & t & \cdots & \frac{t^{n_j-2}}{(n_j-2)!} \\ & & 1 & \ddots & \vdots \\ & & & \ddots & t \\ 0 & & & & 1 \end{bmatrix}.$$

Notice by Remark 6.3.3, we know that $e^{At}P$ is a fundamental solution matrix. That is,

$$e^{At}P = Pe^{Jt}$$

is a fundamental solution matrix. Ideally once we have the Jordan form J, we also have matrix P to obtain the fundamental solution matrix. However, in the practice of solving linear system of differential equations, the matrix P for producing the Jordan form $J = P^{-1}AP$ is not *a priori* known while we know its existence when we have found that A is non-diagonalizable. In the following, we assume that J is known, but P is unknown, and we try to figure how to obtain P.

Let $\{r_{j1}, r_{j2}, \cdots, r_{jn_j}\}$ be the columns of P corresponding to the Jordan block $e^{J_j t}$ in Pe^{Jt}. Then we have

$$A[r_{j1} : r_{j2} : \cdots : r_{jn_j}] = [r_{j1} : r_{j2} : \cdots : r_{jn_j}]J_j.$$

That is,

$$\begin{cases} Ar_{j1} = \lambda_j r_{j1} \\ Ar_{j2} = r_{j1} + \lambda_j r_{j2} \\ Ar_{j3} = r_{j2} + \lambda_j r_{j3} \\ \quad\cdots\cdots \\ Ar_{jn_j} = r_{j(n_j-1)} + \lambda_j r_{jn_j}, \end{cases}$$

which is equivalent to

$$\begin{cases} (A - \lambda_j I)r_{j1} = 0 \\ (A - \lambda_j I)r_{j2} = r_{j1} \\ (A - \lambda_j I)r_{j3} = r_{j2} \\ \quad\cdots\cdots \\ (A - \lambda_j I)r_{jn_j} = r_{j(n_j-1)} \end{cases} \Rightarrow \begin{cases} (A - \lambda_j I)r_{j1} = 0 \\ (A - \lambda_j I)^2 r_{j2} = 0 \\ (A - \lambda_j I)^3 r_{j3} = 0 \\ \quad\cdots\cdots \\ (A - \lambda_j I)^{n_j} r_{jn_j} = 0. \end{cases}$$

Then $\{r_{j1}, r_{j2}, \cdots, r_{jn_j}\}$ are all nonzero vectors; otherwise, the above equalities lead to $r_{j1} = 0$ contradicting the assumption that r_{j1} is an eigenvector. Moreover, we have

$$
\begin{cases}
(A - \lambda_j I)^{n_j} r_{jn_j} = 0 \\
\quad r_{j(n_j-1)} = (A - \lambda_j I) r_{jn_j} \\
\quad r_{j(n_j-2)} = (A - \lambda_j I) r_{j(n_j-1)} = (A - \lambda_j I)^2 r_{jn_j} \\
\quad r_{j(n_j-3)} = (A - \lambda_j I) r_{j(n_j-2)} = (A - \lambda_j I)^3 r_{jn_j} \\
\qquad \cdots\cdots \\
\quad r_{j2} = (A - \lambda_j I) r_{j3} = (A - \lambda_j I)^{n_j-2} r_{jn_j} \\
\quad r_{j1} = (A - \lambda_j I) r_{j2} = (A - \lambda_j I)^{n_j-1} r_{jn_j}.
\end{cases}
$$

Now the algorithm for P is clear: First we solve $(A - \lambda_j I)^{n_j} r_{jn_j} = 0$ for a nonzero vector r_{jn_j}. Then obtain the set $\{r_{j1}, r_{j2}, \cdots, r_{jn_j}\}$ backward in order by multiplying $(A - \lambda_j I)^s$, $s = 1, 2, \cdots, n_j - 1$ on r_{jn_j}, to obtain r_{jn_j-1}, $r_{jn_j-2}, \cdots, r_{j1}$.

Note that $(A - \lambda_j I)^{n_j} r_{jn_j} = 0$ has at least one nonzero solution r_{jn_j} since $\det((A - \lambda_j I)) = 0$. One can show that $\{r_{j1}, r_{j2}, \cdots, r_{jn_j}\}$ determined by the above equalities is linearly independent (see exercise). Then by Lemma 6.2.3, the matrix P so obtained is invertible.

The columns of Pe^{Jt} corresponding to the block for λ_j are then

$$
e^{\lambda_j t} r_{j1};\ e^{\lambda_j t} (t r_{j1} + r_{j2});\ \cdots;\ e^{\lambda_j t} \left(\frac{t^{n_j-1}}{(n_j-1)!} r_{j1} + \frac{t^{n_j-2}}{(n_j-2)!} r_{j2} + \cdots + r_{jn_j} \right).
$$

Example 6.3.9. Find the general solution of $\dot{x} = Ax$ with $A = \begin{bmatrix} 3 & 1 & 0 \\ -4 & -1 & 0 \\ 4 & -8 & -2 \end{bmatrix}$. Then $\det(A - \lambda I) = -(\lambda+2)(\lambda-1)^2$ which has eigenvalues $\lambda_1 = -2$ and $\lambda_2 = 1$ with algebraic multiplicity 2.

For $\lambda_1 = -2$, solving $(A - \lambda_1 I)x = 0$ we obtain an eigenvector

$$
r_1 = \begin{bmatrix} 0 \\ 0 \\ 1 \end{bmatrix}.
$$

For $\lambda_2 = 1$, we solve $(A - \lambda_2 I)^2 x = 0$ to obtain

$$
r_{22} = \begin{bmatrix} -11 \\ -7 \\ 0 \end{bmatrix}, \quad r_{20} = \begin{bmatrix} -3 \\ -6 \\ 20 \end{bmatrix}.
$$

We note that

$$
r_{21} = (A - \lambda_2 I) r_{22} = \begin{bmatrix} 15 \\ -30 \\ 100 \end{bmatrix} \neq 0, \quad \text{but} (A - \lambda_2 I) r_{20} = \begin{bmatrix} 0 \\ 0 \\ 0 \end{bmatrix}.
$$

Therefore $e^{\lambda_2}r_{20}$ can become an independent column of the fundamental solution matrix. But according our algorithm, we use r_{22} instead, which can produce another independent vector r_{21}. Then the matrix P is

$$\left[e^{\lambda_1 t}r_1 \quad \vdots \quad e^{\lambda_2 t}r_{21} \quad \vdots \quad e^{\lambda_2 t}(tr_{21}+r_{22}) \right] = \begin{bmatrix} 0 & 15e^t & (11+15t)e^t \\ 0 & -30e^t & (-7-30t)e^t \\ e^{-2t} & 100e^t & 100te^t \end{bmatrix}.$$

Exercise 6.3.10.

1. Solve the following systems of linear equations.

1)

$$\begin{cases} \dot{x}_1 = x_1 + x_2, \\ \dot{x}_2 = x_1 - x_2. \end{cases}$$

2)

$$\begin{cases} \dot{x}_1 = x_1 + x_2, \\ \dot{x}_2 = x_1 + x_2. \end{cases}$$

3)

$$\begin{cases} \dot{x}_1 = -x_1 + x_2, \\ \dot{x}_2 = x_2 - x_3, \\ \dot{x}_3 = x_1 - 4x_3. \end{cases}$$

4)

$$\begin{cases} \dot{x}_1 = 2x_1 + 2x_2, \\ \dot{x}_2 = -x_2 + x_3, \\ \dot{x}_3 = 2x_3. \end{cases}$$

2. Use change of variables to transform the following different equations into systems of differential equations, and then solve them.

i) $x'' - x' - x = 0$.

ii) $x''' - x'' - 2x' = 0$.

iii) $x'' + 3x' + 2x = 0$.

iv) $x''' + 3x'' + 2x' = 0$.

3. Solve the following system of differential equations

$$\begin{cases} \dot{x}_1 = x_1 + x_2, \\ \dot{x}_2 = -x_1 + x_2. \end{cases}$$

4. Solve the following system of differential equations

$$\begin{cases} \dot{x}_1 = \alpha x_1 - \beta x_2, \\ \dot{x}_2 = \beta x_1 + \alpha x_2, \end{cases}$$

where α, $\beta \in \mathbb{R}$ with $\beta \neq 0$.

5. Solve the following system of differential equations

$$\begin{cases} \dot{x}_1 = x_1 + x_2, \\ \dot{x}_2 = x_2 + x_3, \\ \dot{x}_3 = x_3. \end{cases}$$

6. Let $v \neq 0$ be a nonzero vector in $\mathbb{C}^n$ and $B \in M_{nn}$. If there exists $k \in \mathbb{N}$ such that $B^k v = 0$ but $B^j v \neq 0$ for $1 \leq j < k$, then $\{v, Bv, B^2 v, \cdots, B^{k-1} v\}$ is linearly independent.

6.4 Symmetric matrices and quadratic forms

Example 6.4.1. Let $p(x) = 2x_1^2 + 2x_2^2 + 2x_1 x_2$ be a real-valued polynomial of $x = (x_1, x_2) \in \mathbb{R}^2$. Then $p(x)$ can be rewritten as

$$p(x) = x^T A x = [x_1 \; x_2] \begin{bmatrix} 2 & 1 \\ 1 & 2 \end{bmatrix} \begin{bmatrix} x_1 \\ x_2 \end{bmatrix},$$

where $A = \begin{bmatrix} 2 & 1 \\ 1 & 2 \end{bmatrix}$ is a symmetric real matrix. In order to remove the mixed term $x_1 x_2$, we may find a change of variables $x = Py$ with P invertible such that

$$p(x) = (Py)^T A P y = y^T (P^T A P) y,$$

and $P^T A P = \mathrm{diag}(\lambda_1, \lambda_2)$ is diagonal. Namely

$$p(Py) = y^T \mathrm{diag}(\lambda_1, \lambda_2) y = \lambda_1 y_1^2 + \lambda_2 y_2^2.$$

Indeed, if $P = \begin{bmatrix} \frac{\sqrt{2}}{2} & -\frac{\sqrt{2}}{2} \\ \frac{\sqrt{2}}{2} & \frac{\sqrt{2}}{2} \end{bmatrix}$, then the change of variables $x = Py$ leads to

$$p(x) = p(Py) = y^T (P^T A P) y = [y_1 \; y_2] \begin{bmatrix} 3 & 0 \\ 0 & 1 \end{bmatrix} \begin{bmatrix} y_1 \\ y_2 \end{bmatrix}.$$

That is,

$$p(Py) = y^T \mathrm{diag}(3, 1) y = 3y_1^2 + y_2^2.$$

$\square$

From the above example, it is natural to ask whether or not for every real quadratic polynomial in $x \in \mathbb{R}^n$ there exists a change of variable $x = Py$ such that $p(x) = x^T A x$ can be rewritten

$$p(Py) = y^T \text{diag}(\lambda_1, \lambda_2, \cdots, \lambda_n)y,$$

with P an invertible matrix and $P^T A P = \text{diag}(\lambda_1, \lambda_2, \cdots, \lambda_n)$. That is, we find P such that $P^T A P$ is diagonal. Recall that A is diagonalizable if there is an invertible matrix P such that $P^{-1}AP$ is diagonal and that $P^T = P^{-1}$ if P is an orthogonal matrix.

> **Definition 6.4.2.** A matrix A is called orthogonally diagonalizable if there is an orthogonal matrix P and a diagonal matrix D for which $P^T A P = D$.

Now we are interested in what kind of matrices are orthogonally diagonalizable. First we have

> **Theorem 6.4.3.** If A is orthogonally diagonalizable, then it is symmetric.

Proof: Since A is orthogonally diagonalizable, there is an orthogonal matrix P and a diagonal matrix D for which $P^T A P = D$. Then we have $A = (P^T)^{-1}DP^{-1} = PDP^T$ since $P^T = P^{-1}$. Therefore,

$$A^T = (PDP^T)^T = PDP^T = A.$$

A is symmetric. $\qquad \square$

Next we are curious whether every real symmetric matrix is orthogonally diagonalizable. It turns out to be true. But we need some preparations.

> **Theorem 6.4.4.** If A is a real symmetric matrix, it has n real eigenvalues. That is, every eigenvalue of A is real.

Proof: Suppose that $Ax = \lambda x$. Then we have $\overline{Ax} = \overline{\lambda x}$ which leads to $A\overline{x} = \overline{\lambda}\overline{x}$. Taking transposes on both sides we obtain

$$\overline{x}^T A = \overline{\lambda}\overline{x}^T \Rightarrow \overline{x}^T A x = \overline{\lambda}\overline{x}^T x.$$

Multiplying $\overline{x}^T$ on both sides of $Ax = \lambda x$, we obtain that

$$\overline{x}^T A x = \lambda \overline{x}^T x.$$

Then we have

$$\overline{x}^T A x = \lambda \overline{x}^T x = \overline{\lambda}\overline{x}^T x,$$

leading to $(\overline{\lambda} - \lambda)\overline{x}^T x = 0$. Hence $\overline{\lambda} = \lambda$. λ is real. $\qquad \square$

We leave it as an exercise to show that

Theorem 6.4.5. If A is a real symmetric matrix, then every eigenvectors of A are perpendicular. That is, eigenvectors corresponding to different eigenvalues are orthogonal.

Theorem 6.4.6. A real matrix A is orthogonally diagonalizable if and only if A is symmetric.

Proof: "$\Longrightarrow$": Done with Theorem 6.4.3.

"$\Longleftarrow$": We use mathematical induction on the size of the $n \times n$ matrix A. For $n = 1$, A is a single element matrix, say $A = [a]$. Let $Q = [1]$. Then $Q^T A Q = [1]^T [a][1] = [a]$ is diagonal and $Q = [1]$ is an orthogonal matrix.

Now we consider the general case n and for an inductive assumption, we suppose that every $(n-1) \times (n-1)$ matrix is orthogonally diagonalizable.

Let λ_1 be a real eigenvalue of A and v_1 a real unit eigenvector of A. We can extend v_1 into an orthonormal basis $\mathcal{B}$ of $\mathbb{R}^n$. Group the basis into a matrix $U = [v_1 : v_2 : \cdots : v_n]$. Then U is an orthogonal matrix with $U^T = U^{-1}$ and $U^{-1}AU$ is symmetric. Moreover, the first column of $U^{-1}AU$ is

$$U^{-1}AUe_1 = U^{-1}Av_1 = \lambda_1 U^{-1}v_1 = \lambda_1 e_1,$$

where e_1 is the first vector of the standard basis $(e_1, e_2, \cdots, e_n)$. Therefore, we have

$$U^{-1}AU = \begin{bmatrix} \lambda_1 & 0 \\ 0 & C \end{bmatrix}$$

where C is an $(n-1)\times(n-1)$ symmetric real matrix. By inductive assumption, C is orthogonally diagonalizable. That is, there exists an $(n-1) \times (n-1)$ orthogonal matrix P such that

$$P^T C P = \mathrm{diag}(\lambda_2, \lambda_3, \cdots, \lambda_n).$$

Let $V = \begin{bmatrix} 1 & 0 \\ 0 & P \end{bmatrix}$. Then V is orthogonal and

$$V^{-1}U^{-1}AUV = \begin{bmatrix} 1 & 0 \\ 0 & Q^{-1} \end{bmatrix} \begin{bmatrix} \lambda_1 & 0 \\ 0 & C \end{bmatrix} \begin{bmatrix} 1 & 0 \\ 0 & Q \end{bmatrix}$$
$$= \mathrm{diag}(\lambda_1, \lambda_2, \lambda_3, \cdots, \lambda_n).$$

Let $Q = UV$. Then Q is an orthogonal matrix such that $Q^T AQ$ is diagonal. $\square$

An immediate consequence of Theorem 6.4.6 is the following **spectral decomposition** theorem:

Theorem 6.4.7. If A is an $n \times n$ real symmetric matrix A, then A has an orthogonal eigenvector matrix $Q = [q_1 : q_2 : \cdots : q_n]$ such that

$$A = Q\Lambda Q^T = \lambda_1 q_1 q_1^T + \lambda_2 q_2 q_2^T + \cdots + \lambda_n q_n q_n^T,$$

where $\Lambda = \text{diag}(\lambda_1, \lambda_2, \lambda_3, \cdots, \lambda_n)$ and the main diagonals are eigenvalues of A.

Example 6.4.8. Let $A = \begin{bmatrix} 2 & -2 & 0 \\ -2 & 1 & -2 \\ 0 & -2 & 0 \end{bmatrix}$. Find a spectral decomposition of A.

Solution: Consider $\det(A - \lambda I) = 0$. We have $\det(A - \lambda I) = -(\lambda - 1)(\lambda - 4)(\lambda + 2) = 0$ and the eigenvalues $\lambda_1 = 1$, $\lambda_2 = 4$, $\lambda_3 = -2$. Solving $(A - \lambda_1 I)x = 0$ we have a unit eigenvector,

$$q_1 = \begin{bmatrix} -\frac{2}{3} \\ -\frac{1}{3} \\ \frac{2}{3} \end{bmatrix}.$$

Similarly, solving $(A - \lambda_2 I)x = 0$ and $(A - \lambda_3 I)x = 0$, we have unit eigenvectors corresponding to eigenvalues $\lambda_2 = 4$ and $\lambda_3 = -2$, respectively:

$$q_2 = \begin{bmatrix} \frac{2}{3} \\ -\frac{2}{3} \\ \frac{1}{3} \end{bmatrix}, \ q_3 = \begin{bmatrix} \frac{1}{3} \\ \frac{2}{3} \\ \frac{2}{3} \end{bmatrix}.$$

Then we have an orthogonal matrix,

$$Q = \frac{1}{3} \begin{bmatrix} -2 & 2 & 1 \\ -1 & -2 & 2 \\ 2 & 1 & 2 \end{bmatrix},$$

such that

$$Q^T A Q = \begin{bmatrix} 1 & 0 & 0 \\ 0 & 4 & 0 \\ 0 & 0 & -2 \end{bmatrix}.$$

Moreover, we have

$$A = Q\Lambda Q^T = \lambda_1 q_1 q_1^T + \lambda_2 q_2 q_2^T + \lambda_3 q_3 q_3^T$$

$$= \frac{1}{9} \begin{bmatrix} 4 & 2 & -4 \\ 2 & 1 & -2 \\ -4 & -2 & 4 \end{bmatrix} + \frac{16}{9} \begin{bmatrix} 4 & -4 & 2 \\ -4 & 4 & -2 \\ -2 & -2 & 1 \end{bmatrix} + \frac{-2}{9} \begin{bmatrix} 1 & 2 & 2 \\ 2 & 4 & 4 \\ 2 & 4 & 4 \end{bmatrix}.$$

$\square$

Example 6.4.9. Let $f(x) = x_1^2 + 2x_2^2 + 3x_3^3 - 4x_1x_2 - 4x_2x_3$ with $x = (x_1, x_2, x_3) \in \mathbb{R}^3$. Find an orthogonal transformation $x = Py$ so that $f(Py)$ has no mixed terms of y.

Solution: $f(x) = x^T A x$ with

$$A = \begin{bmatrix} 1 & -2 & 0 \\ -2 & 2 & -2 \\ 0 & -2 & 3 \end{bmatrix}.$$

The characteristic polynomial of A is $\det(A - \lambda I) = -(\lambda - 2)(\lambda + 1)(\lambda - 5)$ with eigenvalues $\lambda_1 = 2$, $\lambda_2 = 5$, $\lambda_3 = -1$. Solving $(A - \lambda I)x = 0$ for each of the eigenvalues, we obtain the following unit eigenvectors:

$$p_1 = \begin{bmatrix} \frac{2}{3} \\ -\frac{1}{3} \\ -\frac{2}{3} \end{bmatrix}, p_2 = \begin{bmatrix} \frac{1}{3} \\ -\frac{2}{3} \\ \frac{2}{3} \end{bmatrix}, p_3 = \begin{bmatrix} \frac{2}{3} \\ \frac{2}{3} \\ \frac{1}{3} \end{bmatrix}.$$

Then we have the orthogonal matrix P given by

$$P = [p_1 : p_2 : p_3] = \frac{1}{3} \begin{bmatrix} 2 & 1 & 2 \\ -1 & -2 & 2 \\ -2 & 2 & 1 \end{bmatrix}.$$

Then we have

$$f(Py) = y^T P^T A P y = y^T \begin{bmatrix} 2 & 0 & 0 \\ 0 & 5 & 0 \\ 0 & 0 & -1 \end{bmatrix} y = 2y_1^2 + 7y_2^2 - y_3^2.$$

Exercise 6.4.10.

1. Find a real symmetric matrix A such that the following polynomials can be represented as $x^T A x$, $x = (x_1, x_2, \cdots, x_n) \in \mathbb{R}^n$, and find a change of variables $x = Py$ with $y = (y_1, y_2, \cdots, y_n) \in \mathbb{R}^n$ such that $f(Py)$ contains no mixed terms $y_i y_j$, $i \neq j$ from y.

1) $f(x) = x_1^2 + x_2^2 + x_1 x_2$.

2) $f(x) = x_1^2 + x_2^2 + 2x_1 x_2$.

3) $f(x) = x_1^2 + x_2^2 + x_3^2 + x_1 x_2 + x_1 x_3 + x_2 x_3$.

4) $f(x) = x_1^2 + x_2^2 + x_3^2 + 2x_1 x_2 + 2x_1 x_3 + 2x_2 x_3$.

5) $f(x) = x_1^2 + x_2^2 + x_1 x_4$.

6) $f(x) = -x_1^2 - x_2^2 + x_1 x_4 + 2x_2 x_3$.

7) $f(x) = x_1^2 + x_2^2 - x_3^2 + x_4^2$.

8) $f(x) = \sum_{i=1}^{n} x_i^2 + \sum_{i=1}^{n-1} x_i x_{i+1}$.

2. Let $p(x) = x^T A x$ be a quadratic polynomial of $x \in \mathbb{R}^3$ with

$$A = \begin{bmatrix} 2 & 2 & -2 \\ 2 & 5 & -4 \\ -2 & -4 & 5 \end{bmatrix}.$$

Find a change of variables $x = Py$ with $y = (y_1, y_2, y_3) \in \mathbb{R}^3$ such that

$$p(Py) = \lambda_1 y_1^2 + \lambda_2 y_2^2 + \lambda_3 y_3^2,$$

for some $\lambda_i \in \mathbb{R}$, $i = 1, 2, 3$.

3. Find a spectral decomposition of each of the following matrices.

$$A = \begin{bmatrix} 2 & 2 & 2 \\ 2 & 2 & 2 \\ 2 & 2 & 2 \end{bmatrix}, \quad B = \begin{bmatrix} 0 & 0 & 2 \\ 0 & 2 & 0 \\ 2 & 0 & 0 \end{bmatrix}.$$

4. Find a spectral decomposition of each of the following matrices.

$$A = \begin{bmatrix} 0 & 0 & 2 & 2 \\ 0 & 2 & 2 & 0 \\ 2 & 2 & 0 & 0 \\ 2 & 0 & 0 & 0 \end{bmatrix}, \quad B = \begin{bmatrix} 0 & 0 & 2 & 2 \\ 0 & 2 & 2 & 2 \\ 2 & 2 & 2 & 0 \\ 2 & 2 & 0 & 0 \end{bmatrix}.$$

5. Let $u = \begin{bmatrix} 1 & 1 & \cdots & 1 \end{bmatrix}$ be a $1 \times n$ matrix and $A = I + u^T u$. Find a spectral decomposition of A.

6. Prove Theorem 6.4.5.

7. Let A be a real symmetric matrix with rank k. Show that A can be written as the sum of k symmetric matrices with rank 1.

8. Let A be a 2×2 real symmetric matrix with eigenvalues $\lambda = \pm 1$. Show that A is an orthogonal matrix.

9. Let A be an $n \times n$ skew-symmetric real matrix. Show that every nonzero eigenvalue of A is purely imaginary.

10. Let A and B be real symmetric matrices. Show that A is similar to B if and only if A and B have the same set of eigenvalues.

11. Let A be an invertible symmetric real matrix and B a skew-symmetric real matrix. Show that if $AB = BA$, then

i) $A + B$ and $A - B$ are invertible;

ii) $(A + B)^{-1}(A - B)$ and $(A - B)^{-1}(A + B)$ are orthogonal matrices.

12. (Schur factorization). Let A be an $n \times n$ real matrix with n real eigenvalues. Show that there exists an orthogonal matrix Q such that $Q^T A Q$ is upper triangular. (Hint: Use mathematical induction.)

6.5   Positive definite matrices

We have confirmed that every real symmetric matrix is orthogonally diagonalizable. Therefore, for every quadratic polynomial $p(x) = x^T A x$ with A a real symmetric matrix, there exists a change of variables $x = Py$ where P is an orthogonal matrix such that

$$p(Py) = y^T P^T A P y = y^T \text{diag}(\lambda_1, \lambda_2, \lambda_3, \cdots, \lambda_n) y = \lambda_1 y_1^2 + \lambda_2 y_2^2 + \cdots + \lambda_n y_n^2,$$

where $\lambda_1, \lambda_2, \lambda_3, \cdots, \lambda_n$ are eigenvalues of A. It follows that if every eigenvalue of A is positive, then for every $x \neq 0$, $y = P^T x \neq 0$, hence

$$p(x) = x^T A x = p(Py) = \lambda_1 y_1^2 + \lambda_2 y_2^2 + \cdots + \lambda_n y_n^2 > 0.$$

Conversely, if

$$p(x) = x^T A x > 0,$$

for every $x \neq 0$, let $A = P^T \text{diag}(\lambda_1, \lambda_2, \lambda_3, \cdots, \lambda_n) P$ with P an orthogonal matrix. Then

$$p(x) = p(Py) = \lambda_1 y_1^2 + \lambda_2 y_2^2 + \cdots + \lambda_n y_n^2 > 0$$

for every $y \neq 0$. It follows that every eigenvalue of A is positive.

Definition 6.5.1. A real symmetric matrix A is said to be **positive definite** if $x^T A x > 0$ for every $x \neq 0$.

We have shown that

Theorem 6.5.2. Let A be a real symmetric matrix A. Then A is positive definite if and only if every eigenvalue of A is positive.

However, it can be computationally inefficient to work out all eigenvalues in order to confirm positive definiteness of a matrix. We have the following equivalent conditions:

Theorem 6.5.3. Let A be a real symmetric $n \times n$ matrix A. Then the following are equivalent.

i) A is positive definite.

ii) Every eigenvalue of A is positive.

iii) $A = B^T B$ for a matrix B with independent columns.

iv) All pivots obtained without row exchange or scalar multiplication of A are positive.

v) All n upper left determinants are positive.

Proof: 1) We have shown i)$\Leftrightarrow$ ii) with Theorem 6.5.2.

2) Next we show ii)$\Leftrightarrow$ iii). "ii)$\Rightarrow$iii)" Let $A = Q\Lambda Q^T$, where $\Lambda = \text{diag}(\lambda_1, \lambda_2, \lambda_3, \cdots, \lambda_n)$ and Q is an orthogonal matrix. Let $\Lambda^{\frac{1}{2}} = \text{diag}(\lambda_1^{\frac{1}{2}}, \lambda_2^{\frac{1}{2}}, \lambda_3^{\frac{1}{2}}, \cdots, \lambda_n^{\frac{1}{2}})$ and $B = \Lambda^{\frac{1}{2}}Q^T$. Then we have $A = B^T B$ and columns of B are linearly independent since $\det(B) > 0$.

"ii)$\Leftarrow$iii)" For every $x \neq 0$, then $Bx \neq 0$ since columns of B are linearly independent. Then we have

$$x^T A x = x^T B^T B x = (Bx)^T Bx > 0.$$

That is, A is positive definite. By Theorem 6.5.2, every eigenvalue of A is positive.

3) Then we show iii)$\Leftrightarrow$iv).

"iii)$\Rightarrow$iv)"

Let $B = QR$ be the QR decomposition of B with Q an orthogonal matrix and R an upper triangular matrix. We can manage to make the main diagonals of R positive by multiplying the corresponding columns of Q by (-1) if necessary. Then we have $A = B^T B = R^T Q^T Q R = R^T R$. That is, A can be written as a product of a lower triangular matrix and an upper triangular matrix with the main diagonal positive. (We call such a decomposition **Cholesky decomposition**.) Therefore, all pivots of A are positive.

"iii)$\Leftarrow$iv)" If all pivots of A are positive, then by LU decomposition (see Section 2.5), there exist an invertible matrix L and a diagonal matrix D with the main diagonal positive such that

$$A = LDL^T.$$

Let $B = L^T D^{\frac{1}{2}}$. We have

$$A = B^T B,$$

and columns of B are linearly independent.

4) Lastly we show v)$\Leftrightarrow$i). "i)$\Rightarrow$v)": Denote by A_k, $k = 1, 2, \cdots, n$ the upper left submatrices of A. Since A is positive definite, for every $x \neq 0$ we have $x^T A x > 0$. Then for every $x = (x_1, x_2, \cdots, x_k, 0, \cdots, 0) \neq 0$ we have

$$x^T A x = [x_1 : x_2 : \cdots : x_k] A_k \begin{bmatrix} x_1 \\ x_2 \\ \vdots \\ x_k \end{bmatrix} > 0.$$

That is, every A_k, $k = 1, 2, \cdots, n$ is positive definite. Then by equivalence of i) and ii) every eigenvalue of A_k is positive. Hence $\det(A_k) > 0$.

"i)$\Leftarrow$v)": Since $\det(A_1) > 0$, we have the first pivot $p_1 = a_{11} > 0$. Then there exists a lower triangular matrix E_1 with 1's in the main diagonal such that

$$E_1 A = \begin{bmatrix} p_1 & & & & * \\ 0 & p_2 & & & \\ \vdots & & * & \ddots & \\ 0 & * & & a_{nn} \end{bmatrix}.$$

Letting E_{12} be the 2×2 upper left submatrix of E_1, we have

$$E_{12} A_2 = \begin{bmatrix} p_1 & * \\ 0 & p_2 \end{bmatrix}.$$

Since $\det(E_{12} A_2) = \det(A_2) > 0$, we have the second pivot $p_2 > 0$. Then there exists a lower triangular matrix E_2 with 1's in the main diagonal such that

$$E_2 A = \begin{bmatrix} p_1 & & & & & * \\ 0 & p_2 & & & & \\ 0 & 0 & p_3 & \cdots & & * \\ 0 & 0 & * & \ddots & & \\ \vdots & \vdots & & & \ddots & \\ 0 & 0 & * & \cdots & & a_{nn} \end{bmatrix}.$$

Letting E_{22} be the 3×3 upper left submatrix of E_2, we have

$$E_2^1 A_3 = \begin{bmatrix} p_1 & * & * \\ 0 & p_2 & * \\ 0 & 0 & p_3 \end{bmatrix}.$$

Since $\det(E_{22} A_3) = \det(A_3) > 0$, we have the second pivot $p_3 > 0$. By the same token we can show every pivots p_k, $k = 1, 2, \cdots, n$ are positive. □

Example 6.5.4. Let $A = \begin{bmatrix} 1 & -1 & 1 \\ -1 & 2 & 0 \\ 1 & 0 & 1 \end{bmatrix}$.

To check the positive definiteness of A, if we solve $\det(A - \lambda I) = 0$ we have $\lambda^3 - 4\lambda^2 + 5\lambda - 3 = 0$, the roots of which are not trivial to obtain. However, we can compute the left upper determinants: $\det(A_1) = 1 > 0$, $\det(A_2) = 1 > 0$, $\det(A_3) = \det(A) = 3 > 0$. Then by Theorem 6.5.3 A is positive definite.

Exercise 6.5.5.

1. Let $A = \begin{bmatrix} 0 & 0 & 0 & \sqrt{10} \\ 0 & -1 & 0 & 0 \\ 0 & 0 & -2 & 0 \\ \sqrt{10} & 0 & 0 & -3 \end{bmatrix}$. Find all possible values of $t \in \mathbb{R}$ such that $A + tI$ is positive definite.

2. Let $A = \begin{bmatrix} 0 & 0 & 0 & 2 \\ 0 & -1 & 0 & 0 \\ 0 & 0 & -2 & 0 \\ 2 & 0 & 0 & -3 \end{bmatrix}$. Find all possible values of $t \in \mathbb{R}$ such that $A + tI$ is positive definite.

3. Show that if A and B are positive definite, so is $A + B$.

4. Let A be a symmetric real matrix. Show that if A is positive definite, then A^{-1} is also positive definite.

5. Show that if A and B are similar and A is positive definite, then B is also positive definite.

6. Let A and B be $n \times n$ positive definite matrices. Is it true AB and BA are positive definite? Justify your answer. (Hint: Use 2×2 matrices to construct examples.)

7. Show that A is positive definite if and only if for every $n \in \mathbb{N}$, there exists a positive definite matrix B such that $A = B^n$.

8. Let A be an $n \times n$ real symmetric matrix with eigenvalues $\lambda_1 \leq \lambda_2 \leq \cdots \leq \lambda_n$. Show that for every $x \in \mathbb{R}^n$ we have

$$\lambda_1 x^T x \leq x^T A x \leq \lambda_n x^T x.$$

9. Show that if A is positive definite, then the function $f : \mathbb{R}^n \to \mathbb{R}$ defined by

$$f(x) = x^T A x - x^T b$$

achieves minimum at x with $2Ax = b$. (Hint: Compute $f(y) - f(x)$ for every $y \in \mathbb{R}^n$.)

10. Let A be a real symmetric $n \times n$ matrix A. We call A **positive semidefinite** if $x^T A x \geq 0$ for every $x \neq 0$. Show that every eigenvalue of A is non-negative.

11. Let A be an $n \times n$ matrix. Show that A is a skew-symmetric matrix if and only if $x^T A x = 0$ for every $x \in \mathbb{R}^n$.

12. Let A be an $n \times n$ symmetric matrix. Show that if $x^T A x = 0$ for every $x \in \mathbb{R}^n$, then $A = 0$.

13. Let A be an $n \times n$ symmetric real matrix. Show that there exists $c > 0$ such that

$$|x^T A x| \leq c x^T x,$$

for every $x \in \mathbb{R}^n$.

14. Let A be an $n \times n$ symmetric real matrix. Show that there exists $t_0 \in \mathbb{R}$ such that for every $t > t_0$, $A + tI$ is positive definite.

15. Let A be an $n \times n$ symmetric real matrix. Show that there exists $t_0 > 0$ such that for every $0 < t < t_0$, $tA + I$ is positive definite.

16. Let A be an $n \times n$ symmetric real matrix. Show that if $\det(A) < 0$, then there exists $x \in \mathbb{R}^n$ such that $x^T A x < 0$.

17. Let A be an $n \times n$ symmetric real matrix. Show that A is positive definite if and only if for every $k \in \mathbb{N}$, there exists a positive definite matrix such that $A = B^k$.

18. Let A and B be $n \times n$ symmetric real matrices and B be positive definite. Show that AB may not be symmetric, but all of the eigenvalues of AB are real.

19. Let A be an $n \times n$ real invertible matrix.

i) Show that there exists a positive definite matrix B such that $A^T A = B^2$;

ii) Show that AB^{-1} is an orthogonal matrix;

iii) Show that every real invertible matrix $A = QB$, where Q is orthogonal and B is positive definite.

20. Show that every real invertible matrix $A = BQ$, where B is positive definite and Q is orthogonal.

Chapter 7

Singular value decomposition

7.1   Singular value decomposition

We have learned that every real symmetric square matrix is orthogonally diagonalizable and enjoy the spectral decomposition according to the eigenvalues and eigenvectors. For general nonsymmetric or nonsquare matrices, there is no spectral decomposition anymore. However, making use of the real symmetric matrix $A^T A$, we may develop a generalized spectral theorem in this section.

We begin with the following theorem which reveals important properties shared by A and $A^T A$.

Theorem 7.1.1. Let A be an $m \times n$ matrix. Then we have

i) A and $A^T A$ have the same nullspace.

ii) A and $A^T A$ have the same row space.

iii) A and $A^T A$ have the same rank.

Proof: i) Let $Ax_0 = 0$. Then we have $A^T A x_0 = 0$. Hence $N(A) \subset N(A^T A)$. Conversely if $A^T A x_0 = 0$, we have

$Ax_0 \perp$ every column of A (or every row of A^T);

Ax_0 itself is a linear combination of columns of A (or every row of A^T).

Ax_0 must be orthogonal to itself. Hence $Ax_0 = 0$. That is, $N(A) \supset N(A^T A)$. Therefore we have $N(A) = N(A^T A)$.

ii) By Theorem 4.1.4, we have $\mathbb{R}^n = N(A) \oplus R(A) = N(A^T A) \oplus R(A^T A)$ where the direct sums are orthogonal. By i) we have $N(A) = N(A^T A)$. Therefore, we have $R(A^T A) = R(A)$.

iii) By ii), since $R(A) = R(A^T A)$, A and $A^T A$ have the same rank. $\quad\square$

We know that $A^T A$ is a real symmetric matrix; therefore, it has orthogonal decomposition

$$A^T A = V D V^T,$$

where $D = \text{diag}\{\lambda_1, \lambda_2, \cdots, \lambda_n\}$ is a diagonal matrix whose main diagonal entries are eigenvalues of $A^T A$, and $V = [v_1 : v_2 : \cdots : v_n]$ is an orthogonal matrix the columns of which are the eigenvectors of $A^T A$ corresponding to the eigenvalues $\lambda_1, \lambda_2, \cdots, \lambda_n$, respectively.

Since for every $x \in \mathbb{R}^n$, and for $y = (y_1, y_2, \cdots, y_n) = V^T x$, we have

$$x^T A^T A x = x^T V D V^T x = \lambda_1 y_1^2 + \lambda_2 y_2^2 + \cdots + \lambda_n y_n^2 \geq 0.$$

Then all eigenvalues $\lambda_1, \lambda_2, \cdots, \lambda_n$ are nonnegative; otherwise we can choose y such that $\lambda_1 y_1^2 + \lambda_2 y_2^2 + \cdots + \lambda_n y_n^2 < 0$.

Definition 7.1.2. Let A be an $m \times n$ matrix. Then every eigenvalues $\lambda_1, \lambda_2, \cdots, \lambda_n$ of $A^T A$ are nonnegative. We call the numbers

$$\sigma_1 = \sqrt{\lambda_1}, \ \sigma_2 = \sqrt{\lambda_2}, \cdots, \sigma_n = \sqrt{\lambda_n}$$

the singular values of A.

Assume that the rank of A is k. Then by Theorem 7.1.1, the rank of $A^T A$ is also k. It follows that D also has rank k since D is similar to $A^T A$. That is,

$$D = \text{diag}\{\lambda_1, \lambda_2, \cdots, \lambda_k, 0, \cdots, 0\}.$$

Notice that

$$Av_i \cdot Av_j = v_j^T A^T A v_i = v_j \cdot \lambda_i v_i = \lambda_i v_i \cdot v_j.$$

We have

$$Av_i \cdot Av_j = \lambda_i v_i \cdot v_j = \begin{cases} 0 & \text{if } i \neq j, \\ \lambda_i & \text{if } i = j. \end{cases}$$

Therefore, $\{Av_1, Av_2, \cdots, Av_k\}$ is an orthogonal set of vectors in the column space of A. Normalization of $\{Av_1, Av_2, \cdots, Av_k\}$ leads to

$$u_1 = \frac{Av_1}{\|Av_1\|} = \frac{Av_1}{\sqrt{\lambda_1}}, \ u_2 = \frac{Av_2}{\|Av_2\|} = \frac{Av_2}{\sqrt{\lambda_2}}, \cdots, u_k = \frac{Av_k}{\|Av_k\|} = \frac{Av_k}{\sqrt{\lambda_k}}.$$

That is, $Av_i = \sqrt{\lambda_1} u_i$, $i = 1, 2, \cdots, k$. Since by Theorem 7.1.1, A and $A^T A$ have the same null space, we have that $A^T A v_i = 0$, $i = k+1, k+2, \cdots, n$ and hence $Av_i = 0$ for $i = k+1, k+2, \cdots, n$. Therefore, we have

$$AV = A[v_1 : v_2 : \cdots : v_k : v_{k+1} : \cdots : v_n]$$

$$=[u_1 : u_2 : \cdots : u_k : 0 : \cdots : 0] \begin{bmatrix} \sqrt{\lambda_1} & & & & & & 0 \\ & \sqrt{\lambda_2} & & & & & \\ & & \ddots & & & & \\ & & & \sqrt{\lambda_k} & & & \\ & & & & 0 & & \\ & & & & & \ddots & \\ 0 & & & & & & 0 \end{bmatrix}.$$

Noticing that V is orthogonal, we have

$$A = [u_1 : \cdots : u_k : 0 : \cdots : 0] \begin{bmatrix} \sqrt{\lambda_1} & & & & & & 0 \\ & \sqrt{\lambda_2} & & & & & \\ & & \ddots & & & & \\ & & & \sqrt{\lambda_k} & & & \\ & & & & 0 & & \\ & & & & & \ddots & \\ 0 & & & & & & 0 \end{bmatrix} V^T$$

$$= \sqrt{\lambda_1} u_1 v_1^T + \sqrt{\lambda_2} u_2 v_2^T + \cdots + \sqrt{\lambda_k} u_k v_k^T.$$

If we extend the orthogonal set $\{u_1 : u_2 : \cdots : u_k\}$ into a full orthonormal basis $\{u_1 : u_2 : \cdots : u_k : u_{k+1} : \cdots : u_m\}$ for $\mathbb{R}^m$ in an arbitrary fashion, we have proved the singular value decomposition theorem

---

Theorem 7.1.3. Let A be an $m \times n$ real matrix with rank k. Then there exist an orthogonal $m \times m$ matrix U, an $m \times n$ matrix

$$\Sigma = \begin{bmatrix} \hat{\Sigma} & 0 \\ 0 & 0 \end{bmatrix},$$

where $\hat{\Sigma} = \text{diag}\{ \sqrt{\lambda_1}, \sqrt{\lambda_2}, \cdots, \sqrt{\lambda_k}\}$, and $\lambda_1, \lambda_2, \cdots, \lambda_k$ are positive eigenvalues of $A^T A$, and an $n \times n$ orthogonal matrix $V = [v_1 : v_2 : \cdots : v_n]$, such that

$$A = U \Sigma V^T = \sqrt{\lambda_1} u_1 v_1^T + \sqrt{\lambda_2} u_2 v_2^T + \cdots + \sqrt{\lambda_k} u_k v_k^T,$$

where the columns of $V = [v_1 : v_2 : \cdots : v_n]$ are eigenvectors of $A^T A$ corresponding to the n nonnegative eigenvalues $\lambda_1, \lambda_2, \cdots, \lambda_k, 0, \cdots, 0$, and $U = [u_1 : u_2 : \cdots : u_k : u_{k+1} : \cdots : u_m]$ satisfies

$$u_i = \frac{A v_i}{\sqrt{\lambda_i}}, i = 1, 2, \cdots, k.$$

The decomposition in Theorem 7.1.3 is called the **full singular value**

decomposition. According to the partition of Σ, we can have the following **reduced singular value decomposition**:

Theorem 7.1.4. Let A be an $m \times n$ real matrix with rank k. Then there exist an $m \times k$ matrix $\hat{U}$ with orthonormal columns, an $k \times k$ diagonal matrix $\hat{\Sigma} = \text{diag}\{ \sqrt{\lambda_1}, \sqrt{\lambda_2}, \cdots, \sqrt{\lambda_k}\}$, where $\lambda_1, \lambda_2, \cdots, \lambda_k$ are positive eigenvalues of $A^T A$, and an $n \times k$ matrix $\hat{V} = [v_1 : v_2 : \cdots : v_k]$ with orthonormal columns, such that

$$A = \hat{U}\hat{\Sigma}\hat{V}^T = \sqrt{\lambda_1}u_1 v_1^T + \sqrt{\lambda_2}u_2 v_2^T + \cdots + \sqrt{\lambda_k}u_k v_k^T,$$

where the columns of $\hat{V} = [v_1 : v_2 : \cdots : v_k]$ are eigenvectors of $A^T A$ corresponding to the k positive eigenvalues $\lambda_1, \lambda_2, \cdots, \lambda_k$, and $\hat{U} = [u_1 : u_2 : \cdots : u_k]$ satisfies

$$u_i = \frac{Av_i}{\sqrt{\lambda_i}}, i = 1, 2, \cdots, k.$$

Example 7.1.5. Find a singular value decomposition of the matrix

$$A = \begin{bmatrix} 0 & 1 \\ 1 & 0 \\ 0 & 1 \end{bmatrix}.$$

Solution: We have

$$A^T A = \begin{bmatrix} 1 & 0 \\ 0 & 2 \end{bmatrix},$$

which has eigenvalues $\lambda_1 = 1$, $\lambda_2 = 2$ with corresponding eigenvectors

$$v_1 = \begin{bmatrix} 1 \\ 0 \end{bmatrix}, v_2 = \begin{bmatrix} 0 \\ 1 \end{bmatrix}.$$

These unit eigenvectors form the columns of V:

$$V = [v_1 : v_2] = \begin{bmatrix} 1 & 0 \\ 0 & 1 \end{bmatrix}.$$

The singular values of A are $\sigma_1 = \sqrt{\lambda_1} = 1$, $\sigma_2 = \sqrt{\lambda_2} = \sqrt{2}$. Then we have

$$\Sigma = \begin{bmatrix} \hat{\Sigma} \\ 0 \end{bmatrix} = \begin{bmatrix} 1 & 0 \\ 0 & \sqrt{2} \\ 0 & 0 \end{bmatrix}.$$

To have $U = [u_1 : u_2 : u_3]$, we firstly compute

$$u_1 = \frac{Av_1}{\sqrt{\lambda_1}} = \begin{bmatrix} 0 \\ 1 \\ 0 \end{bmatrix}, u_2 = \frac{Av_2}{\sqrt{\lambda_2}} = \frac{1}{\sqrt{2}}\begin{bmatrix} 1 \\ 0 \\ 1 \end{bmatrix}.$$

To extend $\{u_1, u_2\}$ into an orthonormal basis for $\mathbb{R}^3$ we need a unit vector $u_3 = (x, y, z)$ such that

$$\begin{cases} u_1^T u_3 = 0, \\ u_2^T u_3 = 0, \end{cases}$$

which leads to $y = 0$ and $x + z = 0$. Hence we can choose

$$u_3 = \frac{1}{\sqrt{2}} \begin{bmatrix} 1 \\ 0 \\ -1 \end{bmatrix}.$$

Then we obtain the full singular decomposition of A:

$$\begin{aligned} A = U\Sigma V^T &= \begin{bmatrix} 0 & \frac{1}{\sqrt{2}} & \frac{1}{\sqrt{2}} \\ 1 & 0 & 0 \\ 0 & \frac{1}{\sqrt{2}} & -\frac{1}{\sqrt{2}} \end{bmatrix} \begin{bmatrix} 1 & 0 \\ 0 & \sqrt{2} \\ 0 & 0 \end{bmatrix} \begin{bmatrix} 1 & 0 \\ 0 & 1 \end{bmatrix}^T \\ &= 1 \cdot u_1 v_1^T + \sqrt{2} u_2 v_2^T \\ &= \begin{bmatrix} 0 \\ 1 \\ 0 \end{bmatrix} \begin{bmatrix} 1 & 0 \end{bmatrix} + \sqrt{2} \begin{bmatrix} \frac{1}{\sqrt{2}} \\ 0 \\ \frac{1}{\sqrt{2}} \end{bmatrix} \begin{bmatrix} 0 & 1 \end{bmatrix}. \end{aligned}$$

The reduced singular value decomposition is

$$\begin{aligned} A = \hat{U}\hat{\Sigma}\hat{V}^T &= \begin{bmatrix} 0 & \frac{1}{\sqrt{2}} \\ 1 & 0 \\ 0 & \frac{1}{\sqrt{2}} \end{bmatrix} \begin{bmatrix} 1 & 0 \\ 0 & \sqrt{2} \end{bmatrix} \begin{bmatrix} 1 & 0 \\ 0 & 1 \end{bmatrix}^T \\ &= 1 \cdot u_1 v_1^T + \sqrt{2} u_2 v_2^T \\ &= \begin{bmatrix} 0 \\ 1 \\ 0 \end{bmatrix} \begin{bmatrix} 1 & 0 \end{bmatrix} + \sqrt{2} \begin{bmatrix} \frac{1}{\sqrt{2}} \\ 0 \\ \frac{1}{\sqrt{2}} \end{bmatrix} \begin{bmatrix} 0 & 1 \end{bmatrix}. \end{aligned}$$

$\square$

Remark 7.1.6. An alternate method to compute the orthogonal matrix U for a singular value decomposition of the $m \times n$ matrix $A = U\Sigma V^T$ is based on the observation that

$$A^T A = V\Sigma^T \Sigma V^T, \; AA^T = U\Sigma\Sigma^T U^T,$$

where $\Sigma^T \Sigma$ and $\Sigma\Sigma^T$ share the same set of positive diagonals $\lambda_1, \lambda_2, \cdots, \lambda_k$, $k = \text{rank}(A)$, with possibly a different number of zeros. Namely, U can be obtained by solving for the set of unit orthogonal eigenvectors of AA^T. $\square$

The Rayleigh quotient

Let A be an $n \times n$ real symmetric matrix. The function $r : \mathbb{R}^n \setminus \{0\} \to \mathbb{R}$ defined by

$$r(x) = \frac{x^T A x}{x^T x}$$

is called the **Rayleigh quotient**. We are interested in when r assumes maximum and minimum. Noticing that $x^T x = \|x\|^2$, we need only consider the extreme values on the unit sphere in $\mathbb{R}^n$. Since A is real symmetric, all eigenvalues are real with $\lambda_1 \geq \lambda_2 \geq \cdots \lambda_n$ and there exists an orthogonal matrix $P = [v_1 : v_2 : \cdots : v_n]$ such that

$$P^T A P = \operatorname{diag}\{\lambda_1, \lambda_2, \cdots, \lambda_n\}.$$

Let $x = Py$ where $\|x\| = 1$. We have $\|y\| = 1$ and

$$
\begin{aligned}
r(x) = r(Py) &= \frac{y^T P^T A P y}{y^T P^T P y} \\
&= \frac{y^T P^T A P y}{y^T y} \\
&= \frac{\lambda_1 y_1^2 + \lambda_2 y_2^2 + \cdots + \lambda_n y_n^2}{y_1^2 + y_2^2 + \cdots + y_n^2} \\
&= \lambda_1 y_1^2 + \lambda_2 y_2^2 + \cdots + \lambda_n y_n^2 \\
&\leq \lambda_1 y_1^2 + \lambda_1 y_2^2 + \cdots + \lambda_1 y_n^2 \\
&= \lambda_1,
\end{aligned}
$$

where the maximum λ_1 is achieved at $\mathbf{y} = e_1 = (1, 0, \cdots, 0)$ by $r(P\mathbf{y})$. That is, the maximum λ_1 is achieved by $r(\mathbf{x})$ at $\mathbf{x} = Pe_1 = v_1$, which is the first column of P and is the eigenvector of A corresponding to λ_1. Similarly,

$$
\begin{aligned}
r(x) = r(Py) &= \frac{y^T P^T A P y}{y^T P^T P y} \\
&= \lambda_1 y_1^2 + \lambda_2 y_2^2 + \cdots + \lambda_n y_n^2 \\
&\geq \lambda_n y_1^2 + \lambda_n y_2^2 + \cdots + \lambda_1 y_n^2 \\
&= \lambda_n,
\end{aligned}
$$

where the minimum λ_n is achieved at $\mathbf{y} = e_n = (0, 0, \cdots, 0, 1)$ by $r(P\mathbf{y})$. That is, the minimum λ_n is achieved by $r(\mathbf{x})$ at $\mathbf{x} = Pe_n = v_n$, which is the last column of P and is the eigenvector of A corresponding to λ_n.

We can also check that if x equals the unit eigenvectors v_1 and v_n, corresponding to λ_1 and λ_n, we have

$$r(x_1) = \frac{x_1^T A x_1}{x_1^T x_1} = \lambda_1, \quad r(x_n) = \frac{x_n^T A x_n}{x_n^T x_n} = \lambda_n.$$

That is, the Rayleigh quotient assumes maximum and minimum at the direction of the eigenvectors corresponding to the maximum and minimum eigenvalues.

The next question is, are there any more? Namely, can the Rayleigh quotient assumes maximum and minimum at non-eigenvector directions? The answer is no. Note that for $x = (x_1, x_2, \cdots, x_j, \cdots, x_n)$, we have

$$\frac{\partial r}{\partial x_j} = \frac{(e_j^T A x + x^T A e_j)}{x^T x} - \frac{x^T A x (e_j^T x + x^T e_j)}{(x^T x)^2}$$

$$= \frac{2(Ax)_j}{x^T x} - \frac{x^T A x (2x_j)}{(x^T x)^2}.$$

If $r(x)$ assumes extreme values at x_0, we have $\frac{\mathrm{d}r}{\mathrm{d}x}(x_0) = 0$. That is,

$$\frac{2(Ax)}{x^T x} - \frac{x^T A x (2x)}{(x^T x)^2}\bigg|_{x=x_0} = 0,$$

which is equivalent to

$$Ax_0 = r(x_0)x_0.$$

That is, the Rayleigh quotient assumes extreme values only at the directions of eigenvectors.

Example 7.1.7. Find the maximum and minimum of $f(x) = x_1^2 + 2x_2^2 + 6x_1x_3 + x_3^2$ on the unit sphere $\|x\| = 1$.

Solution: Since $f(x) = \frac{f(x)}{x^T x}$, we can use the results about the Rayleigh quotient to find the maximum and minimum of f. We first write f in quadratic form $f(x) = x^T A x$ with

$$A = \begin{bmatrix} 1 & 0 & 3 \\ 0 & 2 & 0 \\ 3 & 0 & 1 \end{bmatrix}.$$

The characteristic polynomial is $(2 - \lambda)(1 - \lambda)^2 - 9(2 - \lambda)$. The eigenvalues are $\lambda_1 = 4$, $\lambda_2 = 2$ and $\lambda_3 = -2$.

- The unit eigenvector corresponding to $\lambda_1 = 4$ is

$$v_1 = \frac{1}{\sqrt{2}} \begin{bmatrix} 1 \\ 0 \\ 1 \end{bmatrix}.$$

- The unit eigenvector corresponding to $\lambda_3 = -2$ is

$$v_2 = \frac{1}{\sqrt{2}} \begin{bmatrix} 1 \\ 0 \\ -1 \end{bmatrix}.$$

f achieves maximum value $\lambda_1 = 4$ at v_1 and achieves minimum value $\lambda_3 = -2$ at v_2. □

Exercise 7.1.8.

1. Find a singular value decomposition of $A = \begin{bmatrix} 1 & 0 & 1 \\ 0 & 1 & 0 \end{bmatrix}$.

2. Find a singular value decomposition of $A = \begin{bmatrix} 1 & 0 \\ 0 & 1 \\ 1 & 0 \end{bmatrix}$.

3. Let

$$A = \begin{bmatrix} 2 & -2 & 0 \\ -2 & 1 & -2 \\ 0 & -2 & 0 \end{bmatrix}.$$

Find the vectors $x_{\max}$, $x_{\min}$ in

$$\{x \in \mathbb{R}^3 : x^T x = 1\},$$

where $f(x) = x^T A x$ assumes maximum and minimum, respectively.

4. Let

$$A = \begin{bmatrix} 2 & 2 & 2 \\ 2 & 2 & 2 \\ 2 & 2 & 2 \end{bmatrix}.$$

Find the vectors $x_{\max}$, $x_{\min}$ in

$$\{x \in \mathbb{R}^3 : x^T x = 1\},$$

where $f(x) = x^T A x$ assumes maximum and minimum, respectively.

5. Let $u = \begin{bmatrix} 1 & 1 & \cdots & 1 \end{bmatrix}$ be a $1 \times n$ matrix and $A = I + uu^T$. Find the vectors $x_{\max}$, $x_{\min}$ in

$$\{x \in \mathbb{R}^n : x^T x = 1\},$$

where $f(x) = x^T A x$ assumes maximum and minimum, respectively.

6. Show that for every $n \times n$ matrix A, there exists an orthogonal matrix such that

$$A^T A = Q^T A A^T Q.$$

7. Let A be an $n \times n$ real matrix. Let $R : \mathbb{R}^n \to \mathbb{R}$ be defined by

$$R(x) = \frac{\|Ax\|}{\|x\|}.$$

Find the maximum and minimum of R.

8. Let $A = U_k D V_k^T = \sigma_1 u_1 v_1^T + \sigma_2 u_2 v_2^T + \cdots + \sigma_k u_k v_k^T$ be a singular value decomposition of the $m \times n$ real matrix A where the matrices $U_k = [u_1 : u_2 : \cdots : u_k]$ and $V_k = [v_1 : v_2 : \cdots : v_k]$ both have orthonormal columns, and

$$D = \begin{bmatrix} \sigma_1 & 0 & \cdots & 0 \\ 0 & \sigma_2 & \cdots & 0 \\ \vdots & \vdots & \ddots & \vdots \\ 0 & 0 & \cdots & \sigma_k \end{bmatrix}$$

is an $k \times k$ diagonal matrix with $\sigma_1 \geq \sigma_2 \geq \cdots \geq \sigma_k > 0$.

i) Explain that A and U_k have the same column space.

ii) Explain that for every $b \in \mathbb{R}^m$, $\tilde{b} = U_k U_k^T b$ is the orthogonal projection of b onto the column space of A.

iii) Verify that $\tilde{x} = V_k D^{-1} U_k^T b$ is a least squares solution of $Ax = b$.

9. Let A be an $n \times n$ matrix. Show that there exists a positive semidefinite matrix P and an orthogonal matrix Q such that $A = PQ$. We call the $A = PQ$ a **polar decomposition**.

10. Let A be an $n \times n$ positive definite matrix. Show that for every $n \times n$ matrix B, $B^T A B$ and B have the same rank.

7.2 Principal component analysis

Example 7.2.1. Let $X = [\mathbf{x}_1 : \mathbf{x}_2 : \cdots : \mathbf{x}_n]$ be an $m \times n$ matrix with each column an observation vector of a measurement made on an object $\mathbf{x}$. For example, a survey of $n = 1000$ families on the vector $\mathbf{x} = (x_1, x_2, \cdots, x_8)$ where x_i, $i = 1, 2, \cdots, 8$ stand for annual income, annual expenses on cars, computers, food, medicine, insurance, education, entertainment, respectively, may be written as an 8×1000 matrix X. Then row i of X is a set of observations on the variable x_i contained in $\mathbf{x}$.

One may analyze the data contained in X for different purposes. For example, we may investigate how annual income is correlated with annual expenses on insurance and medicine. Certainly, we assume the underlying correlation is linear, so that we may use the notion of linear dependency to study linear dependency among row 1, row 5 and row 6 of X. □

For convenience, we translate the mean of the observed data to have zero mean, by setting

$$\hat{\mathbf{x}}_i = \mathbf{x}_i - M,$$

where $M = \frac{1}{n} \sum_{i=1}^{n} \mathbf{x}_i$ is the mean of the columns of X. This is a analogy of translating a graph to have the center at the origin. Then the columns of

$$\hat{X} = [\hat{\mathbf{x}}_1 : \hat{\mathbf{x}}_2 : \cdots : \hat{\mathbf{x}}_n]$$

have zero mean. We say that $\hat{X}$ is in **mean-deviation form**. Recall that the dot product $x \cdot y$ with $x, y \in \mathbb{R}^n$ is $x \cdot y = \|x\| \cdot \|y\| \cos \theta$, where θ is defined to be the angle between x and y. We observe that

$$\cos \theta = \frac{x}{\|x\|} \cdot \frac{y}{\|y\|}$$

implies that if $\cos \theta = 0$, $x \perp y$ and x, y are linearly independent; if $\cos \theta = 1$, x, y are co-linear and are linearly dependent. Therefore, the dot product $x \cdot y$ may be regarded as a measure of how much x and y are correlated. For instance, if the first row r_1 of X is the annual income, second row r_2 the annual expenses on cars, then $r_1 \cdot r_2$ gives an indication of how annual income is correlated with annual expenses on cars.

We call the matrix,

$$S = \frac{1}{n-1} X X^T,$$

the sample **covariance matrix**, where S_{ij} measures how much the i-th row of $\hat{X}$ are correlated to the j-th row of $\hat{X}$. We call

$$S_{jj} = \frac{1}{n-1} \sum_{i=1}^{n} (X)_{ji} (X^T)_{ij} = \frac{1}{n-1} \sum_{i=1}^{n} (X)_{ji}^2$$

the variance of the j-th variable x_j of the object vector $\mathbf{x}$. The variance of x_j measures the **spread** of the values of x_j around the zero mean.

The trace of S,

$$\text{trace}(S) = \frac{1}{n-1} \sum_{j=1}^{m} S_{jj},$$

is called the **total variance** of the data X.

In this section, we introduce the so-called **principal component analysis**, which is a procedure that uses an orthogonal transformation $\mathbf{x} = P\mathbf{y}$ to convert the observations of $\mathbf{x}$ whose variables are possibly correlated into a set $\mathbf{y}$ of linearly uncorrelated variables. We call the linearly uncorrelated variables of $\mathbf{y}$ the **principal components.**

To wit, we assume that the $m \times n$ matrix $X = [\mathbf{x}_1 : \mathbf{x}_2 : \cdots : \mathbf{x}_n]$ has zero mean and $S = \frac{1}{n-1} X X^T$ is the covariance matrix. We find an orthogonal $m \times m$ matrix P such that the change of variables,

$$\mathbf{x} = P\mathbf{y} \Longleftrightarrow \begin{bmatrix} x_1 \\ x_2 \\ \vdots \\ x_m \end{bmatrix} = P \begin{bmatrix} y_1 \\ y_2 \\ \vdots \\ y_m \end{bmatrix},$$

has the property that the new variables $y_1, y_2, \cdots, y_p$ of $\mathbf{y}$ are uncorrelated. That is, the row of the values for y_i is orthogonal to that for y_j, if $i \neq j$. Let $Y = [\mathbf{y}_1 : \mathbf{y}_2 : \cdots : \mathbf{y}_m] = P^T X$. We want the covariance matrix

$$\frac{1}{n-1} Y Y^T = \frac{1}{n-1}[P^T \mathbf{x}_1 : P^T \mathbf{x}_2 : \cdots : P^T \mathbf{x}_n][P^T \mathbf{x}_1 : P^T \mathbf{x}_2 : \cdots : P^T \mathbf{x}_n]^T$$

$$= \frac{1}{n-1} P^T X X^T P$$

$$= P^T S P$$

to be diagonal. Since $S = \frac{1}{n-1} X X^T$ is real symmetric, such an orthogonal matrix P exists. Moreover, since S is positive semidefinite (i.e., $x^T S x \geq 0$ for every $x \neq 0$, see Exercise 6.5.5), the eigenvalues satisfy $\lambda_1 \geq \lambda_2 \geq \cdots \geq \lambda_m \geq 0$. Let

$$D = \mathrm{diag}\{\lambda_1, \lambda_2, \cdots, \lambda_m\}.$$

Then $P^T S P = D$. That is, $S P = P D$. The unit vectors of columns of $P = [p_1 : p_2 : \cdots : p_m]$ are eigenvectors of the covariance matrix S and are called the (directions of) **principal components** of the data of observations. We call the eigenvector corresponding to the largest eigenvalue the **first principal component**; and call the eigenvector corresponding to the second largest eigenvalue the **second principal component**, and so on. Moreover, $\mathbf{y}$ with

$$y_i = p_i^T \mathbf{x}, i = 1, 2, \cdots, m,$$

becomes the new variable with uncorrelated coordinates and λ_i measures the variance of the new variable y_i.

Notice that by Theorem 6.2.6, $\mathrm{trace}(S) = \mathrm{trace}(P^T S P)$. The total variance of S is not changed after orthogonal diagonalization. Then the fraction

$$\frac{\lambda_i}{\mathrm{trace}(S)} = \frac{\lambda_i}{\lambda_1 + \lambda_2 + \cdots + \lambda_m}$$

indicates the portion of variance contributed by the i-th principal component p_i.

Example 7.2.2. Let the following table list the data of the annual income and annual living expenses of five families (in thousand dollars):

	1	2	3	4	5
income	70	125	125	135	250
living expenses	50	60	61	64	70

i) Find the covariance matrix for the data;

ii) Make a principal component analysis of the data to find all the principal components;

iii) Find the total variance T in the data, and the fraction contributed by the first principal component.

Solution: i) Let X_i, $i = 1, 2, \cdots, 5$ denote the i-th column of the table of the data. The sample mean vector is

$$\mathbf{m} = \frac{1}{5} \sum_{i=1}^{5} X_i = \begin{bmatrix} 141 \\ 61 \end{bmatrix}.$$

Let $A = [X_1 - \mathbf{m} : X_2 - \mathbf{m} : \cdots : X_5 - \mathbf{m}]$. Then

$$A = \begin{bmatrix} -71 & -16 & -16 & 6 & 109 \\ -11 & -1 & 0 & 3 & 9 \end{bmatrix}.$$

The covariance matrix S is then

$$S = \frac{1}{5-1} A A^T$$
$$= \frac{1}{4} \begin{bmatrix} 17470 & 1796 \\ 1796 & 212 \end{bmatrix}.$$

ii) The eigenvalues of S are

$$\lambda_1 \approx 4413.73, \lambda_2 \approx 6.76899.$$

The corresponding vectors are

$$u_1 \approx \begin{bmatrix} -0.994741 \\ -0.102423 \end{bmatrix}, u_2 \approx \begin{bmatrix} 0.102423 \\ -0.994741 \end{bmatrix},$$

where u_1 is the first principal component and u_2 the second principal component.

iii) The total variance T is

$$T = \lambda_1 + \lambda_2 = 4420.5 = \text{trace}(S).$$

The fraction contributed by the first principal component is $\frac{\lambda_1}{T} = \frac{4413.73}{4420.5} \approx$ 99.8468%. □

Exercise 7.2.3.

1. The following table lists the data of the scores of five students:

	1	2	3	4	5
Exam I	70	75	85	95	60
Exam II	50	60	75	65	70
Exam III	76	59	61	80	80

i) Find the covariance matrix for the data;

ii) Make a principal component analysis of the data to find all the principal components;

iii) Find the total variance T in the data, and the fraction contributed by the first principal component.

2. Let $X = [\mathbf{x}_1 : \mathbf{x}_2 : \cdots : \mathbf{x}_n]$ be an $m \times n$ matrix with zero mean and P an $m \times m$ invertible matrix. Show that $Y = [P^T \mathbf{x}_1 : P^T \mathbf{x}_2 : \cdots : P^T \mathbf{x}_n]$ has zero mean.

Chapter 8

Linear transformations

A fundamental theme of many branches of mathematics is the study of functions, or transformations, between vector spaces. A function can be first classified as linear or nonlinear. For example, $f : \mathbb{R} \to \mathbb{R}$ with $f(x) = 2x$ is a linear function, while $g : \mathbb{R} \to \mathbb{R}$ with $g(x) = x^2$ is a nonlinear function. A linear function can be further classified as homogeneous linear or nonhomogeneous linear. For example, $f : \mathbb{R} \to \mathbb{R}$ with $f(x) = 2x$ is homogeneous linear while $h : \mathbb{R} \to \mathbb{R}$ with $h(x) = 2x + 1$ is nonhomogeneous linear. As an important application of matrix theory, we devote this chapter to a brief discussion of homogeneous linear functions between vector spaces. In what follows, if no confusion otherwise arises, we say that a function is linear when it is homogeneous linear.

8.1   Linear transformation and matrix representation

Definition 8.1.1. If $T : V \to W$ is a function where V and W are vector spaces such that for every u, $v \in V$ and scalar k,

$$T(u + v) = T(u) + T(v),$$
$$T(ku) = kT(u),$$

then we call $T : V \to W$ a **linear transformation**. If, in addition, $V = W$ we call T a **linear operator** on V.

Example 8.1.2. Let A be an $m \times n$ real matrix. Then the map $T : \mathbb{R}^n \to \mathbb{R}^m$

defined by $T(x) = Ax$ is a linear transformation, since for every u, $v \in \mathbb{R}^n$ and scalar $k \in \mathbb{R}$,

$$T(u + v) = A(u + v) = Au + Av = T(u) + T(v),$$
$$T(ku) = A(ku) = kAu = kT(u).$$

□

Example 8.1.3. Let $A = [c_1 : c_2 : \cdots : c_n]$ be an $m \times n$ real matrix with rank$(A) = n$. Then the projection $P : \mathbb{R}^n \to \mathbb{R}^m$ is a linear transformation from $\mathbb{R}^n$ to $\mathbb{R}^m$. Indeed, for every $b \in \mathbb{R}^n$, we have the orthogonal projection $Pb = A(A^T A)^{-1} A^T b$. Then for every u, $v \in \mathbb{R}^n$ and scalar $k \in \mathbb{R}$,

$$P(u + v) = A(A^T A)^{-1} A^T (u + v) = Pu + Pv,$$
$$P(ku) = kP(u).$$

P is a linear transformation from $\mathbb{R}^n$ to $\mathbb{R}^m$. □

Example 8.1.4. Let P_n denote the vector space of all real polynomials with degree less than or equal to n. The map $T : P_2 \to P_3$ defined by

$$T(p)(x) = xp(x)$$

is a linear transformation. Indeed, for every f, $g \in P_2$ and $k \in \mathbb{R}$, we have

$$T(f + g)(x) = x(f(x) + g(x)) = xf(x) + xg(x) = T(f)(x) + T(g)(x),$$
$$T(kf)(x) = x(kf(x)) = k(xf(x)).$$

Therefore T is a linear transformation. □

Example 8.1.5. Let V be a real vector space with dim $V = n$ and a basis $\{v_1, v_2, \cdots, v_n\}$. The coordinate map $[\,] : V \to \mathbb{R}^n$ defined by

$$[\,](x) = [x]_V$$

is a linear transformation. Indeed, for every u, $v \in V$ and scalar $k \in \mathbb{R}$, we have

$$
\begin{aligned}
[\,](u + v) &= [u + v]_V \\
&= \left[[v_1 : v_2 : \cdots : v_n][u]_V + [v_1 : v_2 : \cdots : v_n][v]_V\right]_V \\
&= \left[[v_1 : v_2 : \cdots : v_n]([u]_V + [v]_V V)\right]_V \\
&= [u]_V + [v]_V]_V \\
&= [\,](u) + [\,](v), \\
[\,](ku) &= [ku]_V \\
&= \left[k[v_1 : v_2 : \cdots : v_n][u]_V\right]_V \\
&= \left[[v_1 : v_2 : \cdots : v_n]k[u]_V\right]_V
\end{aligned}
$$

$$= k[u]_V$$
$$= k[\ ](u).$$

That is, the coordinate map $[\ ] : V \to \mathbb{R}^n$ is a linear transformation. □

Remark 8.1.6. Recall that a function $f : A \to B$ is said to be **one to one**, or **injective** , if for every $x_1, x_2 \in A$ with $f(x_1) = f(x_2)$ we have $x_1 = x_2$. $f : A \to B$ is said to be **onto**, or **surjective** , if for every $y \in B$ there exists $x \in A$ such that $y = f(x)$. If $f : A \to B$ is both injective and surjective, f is called a **bijection**.

Using the fact that a basis of a vector space V with $\dim V = n$ is linearly independent, we can show that the coordinate map $[\ ] : V \to \mathbb{R}^n$ is a one-to-one and onto map. □

Theorem 8.1.7. Let $T : V \to W$ be a function from a vector space V to a vector space W with $\dim V = n$ and $\dim W = m$. Let $S = \{v_1, v_2, \cdots, v_n\}$ be a basis of V and $L = \{w_1, w_2, \cdots, w_m\}$ be a basis of W. T is a linear transformation if and only if there exists an $m \times n$ matrix A such that for every $x \in V$,

$$[T(x)]_L = A[x]_S.$$

Moreover, if T is a linear transformation, then

$$A = \big[[T(v_1)]_L : [T(v_2)]_L : \cdots : [T(v_n)]_L\big].$$

Proof. "$\Longrightarrow$" Let $\{v_1, v_2, \cdots, v_n\}$ be a basis of V. For every $x \in V$, there exists a coordinate vector $[x]_v = (x_1, x_2, \cdots, x_n) \in \mathbb{R}^n$ such that

$$x = [v_1 : v_2 : \cdots : v_n][x]_S.$$

Then by linearity of A we have

$$
\begin{aligned}
T(x) &= T([v_1 : v_2 : \cdots : v_n][x]_S)\\
&= T(x_1 v_1 + x_2 v_2 + \cdots + x_n v_n)\\
&= x_1 T(v_1) + x_2 T(v_2) + \cdots + x_n T(v_n)\\
&= x_1[w_1 : w_2 : \cdots : w_m][T(v_1)]_L + x_2[w_1 : w_2 : \cdots : w_m][T(v_2)]_L\\
&\quad + \cdots + x_n[w_1 : w_2 : \cdots : w_m][T(v_n)]_L\\
&= [w_1 : w_2 : \cdots : w_m]\,(x_1[T(v_1)]_L + x_2[T(v_2)]_L + \cdots + x_n[T(v_n)]_L)\\
&= [w_1 : w_2 : \cdots : w_m]\big[[T(v_1)]_L : [T(v_2)]_L : \cdots : [T(v_n)]_L\big][x]_S.
\end{aligned}
$$

That is,

$$[T(x)]_L = A[x]_S,$$

where
$$A = [[T(v_1)]_L : [T(v_2)]_L : \cdots : [T(v_n)]_L].$$

"$\Longleftarrow$" If for every $x \in V$,

$$[T(x)]_L = A[x]_S.$$

Then we have

$$\begin{aligned}
T(x) &= [w_1 : w_2 : \cdots : w_m][T(x)]_L \\
&= [w_1 : w_2 : \cdots : w_m]A[x]_S.
\end{aligned}$$

For every $u, v \in V$ and scalar k we have

$$\begin{aligned}
T(u + v) &= [w_1 : w_2 : \cdots : w_m]A[u + v]_S \\
&= [w_1 : w_2 : \cdots : w_m]A([u]_S + [v]_S) \\
&= T(u) + T(v), \\
T(ku) &= [w_1 : w_2 : \cdots : w_m]A[ku]_S \\
&= [w_1 : w_2 : \cdots : w_m]Ak[u]_S \\
&= kT(u).
\end{aligned}$$

That is, T is a linear transformation. $\qquad\qquad\qquad\qquad$ □

Remark 8.1.8. Fix a basis $S = \{v_1, v_2, \cdots, v_n\}$ of V and a basis $L = \{w_1, v = w_2, \cdots, w_m\}$ of W. We call the matrix $A = [[T(v_1)]_L : [T(v_2)]_L : \cdots : [T(v_n)]_L]$ defined in Theorem 8.1.7 the **representation matrix** of the transformation $T : V \to W$ with respect to basis S and L, and denote it by $[T]$ when the bases of V and W are clear.

When $T : V \to W$ is a linear transformation with $V = W$, we call T a **linear operator** on V. In this case we usually choose the same basis S for both of the range and domain of T and use $[T]_S$ to denote the representation matrix of T. $\qquad\qquad\qquad\qquad$ □

Theorem 8.1.7 shows that a linear transformation between finite dimensional vector spaces can be completely described by a matrix multiplication of the coordinate vectors. Note also that the coordinate map $[\ ] : V \to \mathbb{R}^n$ defined by $[\ ](x) = [x]_V$ is a bijection. Thus properties of linear transformations can be investigated by studying the representation matrix.

Example 8.1.9. Let $T : P_2 \to P_3$ be defined by

$$T(p)(x) = xp(x).$$

Then by Example 8.1.4 T is a linear transformation. Let $\{1, x, x^2\}$ be a basis of P_2 and $\{1, x, x^2, x^3\}$ be a basis of P_3. Then the matrix representation matrix A of T is

$$A = [[T(1)]_{P_3} : T(x)]_{P_3} : T(x^2)]_{P_3}]$$

$$= \begin{bmatrix} 0 & 0 & 0 \\ 1 & 0 & 0 \\ 0 & 1 & 0 \\ 0 & 0 & 1 \end{bmatrix}.$$

Let $f(x) = 1 + x^2$. Then we have

$$[T(f)]_{P_3} = A[f]_{P_2}$$

$$= \begin{bmatrix} 0 & 0 & 0 \\ 1 & 0 & 0 \\ 0 & 1 & 0 \\ 0 & 0 & 1 \end{bmatrix} \begin{bmatrix} 1 \\ 0 \\ 1 \end{bmatrix}$$

$$= \begin{bmatrix} 0 \\ 1 \\ 0 \\ 1 \end{bmatrix}.$$

Therefore we have

$$T(f)(x) = \begin{bmatrix} 1 & x & x^2 & x^3 \end{bmatrix} [T(f)]_{P_3}$$

$$= \begin{bmatrix} 1 & x & x^2 & x^3 \end{bmatrix} \begin{bmatrix} 0 \\ 1 \\ 0 \\ 1 \end{bmatrix}$$

$$= x + x^3.$$

□

Exercise 8.1.10.

1. Determine whether each of the following functions from $\mathbb{R}^n \to \mathbb{R}^m$ is a linear transformation.

    i) $f(x) = 1, x \in \mathbb{R}$;

    ii) $f(x) = x, x \in \mathbb{R}$;

    iii) $f(x) = x + 1, x \in \mathbb{R}$;

    iv) $f(x) = 2x, x \in \mathbb{R}$;

    v) $f(x) = x_1 + x_2, x = (x_1, x_2) \in \mathbb{R}^2$;

    vi) $f(x) = (2x_1, x_2), x = (x_1, x_2) \in \mathbb{R}^2$;

    vii) $f(x) = (x_1, x_1), x = (x_1, x_2) \in \mathbb{R}^2$;

   viii) $f(x) = (0, 0), x = (x_1, x_2) \in \mathbb{R}^2$.

2. Let $T : P_2 \to P_2$ be defined by

$$T(p)(x) = p(x) - p(0).$$

i) Show that T is a linear transformation;

ii) Find the representation matrix of T under the standard basis of P_2 and find its rank;

iii) Determine whether or not T is injective and whether or not T is surjective.

3. Let V be a vector space with dim $V = n$. Let $B_1 = \{v_1, v_2, \cdots, v_n\}$ be a basis of V. For every set $\{x_1, x_2, \cdots, x_n\}$ in V, there exists a unique linear transformation $T : V \to V$ such that $T(v_i) = x_i$, $i = 1, 2, \cdots, n$.

4. Let $M_{nn}(R)$ be the vector space of all $n \times n$ real matrices. Let $B \in M_{nn}(R)$ and define $T : M_{nn}(R) \to M_{nn}(R)$ by

$$T(A) = AB - BA.$$

i) Show that T is a linear operator.

ii) Find the matrix representation $[T]$ of T under the standard basis of $M_{nn}(R)$.

iii) Show that for every $A, C \in M_{nn}(R)$, $T(AC) = AT(C) + CT(A)$.

5. Let V be a vector space and $T : V \to V$ a linear operator on V. If $W \subset V$ is a subspace of V, then

$$\dim(TV) + \dim W - \dim(TW) \leq n.$$

6. Let V be a vector space and $T : V \to V$ a linear operator on V. We say that T is an **invertible linear transformation** if there exists an operator S on V such that $TS = ST = I$, where I is the identity map. Let $S = \{\epsilon_1, \epsilon_2, \cdots, \epsilon_n\}$ be a basis of V.

i) Show that T is invertible if and only if $\{T\epsilon_1, T\epsilon_2, \cdots, T\epsilon_n\}$ is linearly independent;

ii) Show that T is invertible if and only if T is one to one;

iii) Show that T is invertible if and only if T is onto.

8.2 Range and null spaces of linear transformation

Definition 8.2.1. Let $T : V \to W$ be a linear transformation where V and W are vector spaces. We call

$$\ker(T) = \{x \in V : Tx = 0\}$$

the **kernel** of T, and call

$$\text{Range}(T) = \{Tx : x \in V\}$$

the **range** of T.

By linearity of T, we have that for every x_1, $x_2 \in \ker(T)$ and for every scalar k_1 and k_2,

$$T(k_1 x_1 + k_2 x_2) = T(k_1 x_1) + T(k_2 x_2) = k_1 T(x_1) + k_2 T(x_2) = 0.$$

That is, $k_1 x_1 + k_2 x_2 \in \ker(T)$. Therefore, $\ker(T)$ is a subspace of V. Similarly, for every y_1, $y_2 \in \text{Range}(T)$ and for every scalar k_1 and k_2, there exists x_1, $x_2 \in V$ such that $T(x_1) = y_1$, $T(x_2) = y_2$ and

$$k_1 y_1 + k_2 y_2 = k_1 T(x_1) + k_2 T(x_2) = T(k_1 x_1 + k_2 x_2).$$

That is, $k_1 x_1 + k_2 x_2 \in \text{Range}(T)$. Therefore, $\text{Range}(T)$ is a subspace of W. We arrive at

---

Theorem 8.2.2. Let $T : V \to W$ be a linear transformation where V and W are vector spaces. Then $\ker(T)$ is a subspace of V and $\text{Range}(T)$ is a subspace of W.

Let $T : V \to W$ be a linear transformation where V and W are vector spaces with $\dim V = n$ and $\dim W = m$. Let $\{v_1, v_2, \cdots, v_n\}$ be a basis of V and $\{w_1, w_2, \cdots, w_m\}$ be a basis of W. Then by Theorem 8.1.7, there exists an $m \times n$ matrix A such that $[T(x)]_W = A[x]_v$. Then we have

$$\begin{aligned}
\ker(T) =& \{x \in V : Tx = 0\} \\
=& \{x \in V : [w_1 : w_2 : \cdots : w_m][Tx]_W = 0\} \\
=& \{x \in V : [Tx]_W = 0\} \\
=& \{x \in V : A[x]_v = 0\}.
\end{aligned}$$

That is,

The kernel of T is the set of all vectors in V whose coordinate vectors are in $N(A)$.

Then by Example 8.1.5 and Remark 8.1.6 that the coordinate map $[\ ]$: $V \to \mathbb{R}^n$ is a bijection we have

$$\dim \ker(T) = \dim N(A).$$

Similarly, we have

$$\text{Range}(T) = \{Tx : x \in V\}$$

$$= \{[w_1 : w_2 : \cdots : w_m][Tx]_W : x \in V\}$$
$$= \{[w_1 : w_2 : \cdots : w_m]A[x]_v : x \in V\}.$$

We have $\dim \mathrm{Range}(T) = \dim C(A)$. That is,

> The range of T is the set of all vectors in W whose coordinate vectors are in $C(A)$.

Recall that for an $m \times n$ real matrix A, we have $\dim N(A) + \dim C(A) = n$. For linear transformations we have the following similar results.

> **Theorem 8.2.3.** Let $T : V \to W$ be a linear transformation where V and W are vector spaces with $\dim V = n$ and $\dim W = m$. Then we have
>
> $$\dim \ker(T) + \dim \mathrm{Range}(T) = n.$$

Example 8.2.4. (Example 8.1.9 revisited) Let $T : P_2 \to P_3$ be defined by

$$T(p)(x) = xp(x).$$

Then the matrix representation matrix A of T is

$$A = [[T(1)]_{P_3} : T(x)]_{P_3} : T(x^2)]_{P_3}]$$
$$= \begin{bmatrix} 0 & 0 & 0 \\ 1 & 0 & 0 \\ 0 & 1 & 0 \\ 0 & 0 & 1 \end{bmatrix}.$$

Then we have $N(A) = \{\mathbf{0}\} \in \mathbb{R}^3$ and

$$\ker(T) = \{\mathbf{0}\} \subset P_2.$$

Moreover, $C(A) = \{(0, x) \in \mathbb{R}^4 : x \in \mathbb{R}^3)$. Therefore we have

$$\mathrm{Range}(T) = \left\{ \begin{bmatrix} 1 & x & x^2 & x^3 \end{bmatrix} \begin{bmatrix} 0 \\ x_1 \\ x_2 \\ x_3 \end{bmatrix} : (x_1, x_2, x_3) \in \mathbb{R}^3 \right\} = \mathrm{span}\{x, x^2, x^3\}.$$

□

> **Theorem 8.2.5.** Let V and W be vector spaces, $T : V \to W$ a linear transformation. Then T is one to one if and only if $\ker(T) = \{\mathbf{0}\}$.

Proof. "$\Longrightarrow$" Suppose $\ker(T) \neq \{\mathbf{0}\}$. Then there exists $x_1 \neq 0$, $x_1 \in V$ such that $T(x_1) = 0 = T(0)$. T is not one to one. This is a contradiction.

"$\Longleftarrow$" Let $T(x_1) = T(x_2)$, x_1, $x_2 \in V$. Then we have $T(x_1 - x_2) = 0$ and $x_1 = x_2$ since $\ker(T) = \{\mathbf{0}\}$. □

By Theorems 8.2.3 and 8.2.5, we have that

Theorem 8.2.6. Let $T : V \to W$ be a linear transformation where V and W are vector spaces with $\dim V = \dim W = n$. Then T is one to one if and only if T is onto.

Let $T : V \to V$ be a linear operator on V. It turns out the representation matrices of T with respect to different bases are similar. To be more specific we have

---

Theorem 8.2.7. Let V be a vector space with $\dim V = n$. Let $T : V \to V$ be a linear transformation. Let $S = \{v_1, v_2, \cdots, v_n\}$ and $L = \{w_1, w_2, \cdots, w_n\}$ be bases of V. Then we have

$$[T]_S = P[T]_L P^{-1},$$

where P is the transition matrix from basis L to S.

Proof. For every $x \in V$, we have

$$
\begin{aligned}
Tx &= [v_1 : v_2 : \cdots : v_n][Tx]_S \\
&= [v_1 : v_2 : \cdots : v_n][T]_S[x]_S \\
&= [w_1 : w_2 : \cdots : w_n]P^{-1}[T]_S P[x]_L.
\end{aligned}
$$

We also have

$$Tx = [w_1 : w_2 : \cdots : w_n][Tx]_L = [w_1 : w_2 : \cdots : w_n][T]_L[x]_L.$$

Then we have

$$P^{-1}[T]_S P[x]_L = [T]_L[x]_L.$$

Since x is arbitrary, we have $[T]_L = P^{-1}[T]_S P$. □

Example 8.2.8. Let $T : \mathbb{R}^3 \to \mathbb{R}^3$ be a linear operator which has the representation matrix

$$[T]_L = \begin{bmatrix} 1 & 0 & -1 \\ 1 & 1 & 0 \\ -1 & 2 & 1 \end{bmatrix},$$

under the basis $L = \{\eta_1, \eta_2, \eta_3\}$ where

$$\eta_1 = \begin{bmatrix} -1 \\ 0 \\ 1 \end{bmatrix}, \eta_2 = \begin{bmatrix} 0 \\ -1 \\ 1 \end{bmatrix}, \eta_3 = \begin{bmatrix} -1 \\ 1 \\ 0 \end{bmatrix}.$$

Find the representation matrix $[T]_S$, where $S = \{\epsilon_1, \epsilon_2, \epsilon_3\}$ is the standard basis of $\mathbb{R}^3$.

Solution: The transition matrix from L to S is

$$Q = \begin{bmatrix} \eta_1 : \eta_2 : \eta_3 \end{bmatrix}$$

$$= \begin{bmatrix} -1 & 0 & -1 \\ 0 & -1 & 1 \\ 1 & 1 & 0 \end{bmatrix}.$$

That is, $[\eta_1 : \eta_2 : \eta_3] = [\epsilon_1 : \epsilon_2 : \epsilon_3]Q$. Then the transition matrix from S to L is $Q^{-1} = \frac{1}{3}\begin{bmatrix} -1 & 0 & 1 \\ 0 & -1 & 1 \\ -1 & 1 & 0 \end{bmatrix}$. Therefore we have

$$[T]_S = Q[T]_L Q^{-1}$$

$$= \frac{1}{3}\begin{bmatrix} -1 & 0 & -1 \\ 0 & -1 & 1 \\ 1 & 1 & 0 \end{bmatrix}\begin{bmatrix} 1 & 0 & -1 \\ 1 & 1 & 0 \\ -1 & 2 & 1 \end{bmatrix}\begin{bmatrix} -1 & 0 & 1 \\ 0 & -1 & 1 \\ -1 & 1 & 0 \end{bmatrix}$$

$$= \frac{1}{3}\begin{bmatrix} 0 & 2 & -2 \\ 1 & 0 & -1 \\ -1 & -2 & 3 \end{bmatrix}.$$

That is,

$$T[\epsilon_1 : \epsilon_2 : \epsilon_3] = [\epsilon_1 : \epsilon_2 : \epsilon_3][T]_S.$$

$\square$

Exercise 8.2.9.

1. Let $T : \mathbb{R}^3 \to \mathbb{R}^3$ be a linear operator which has the representation matrix

$$[T]_L = \begin{bmatrix} -4 & 0 & -4 \\ 0 & -1 & -1 \\ 3 & 6 & 9 \end{bmatrix},$$

under the basis $L = \{\eta_1, \eta_2, \eta_3\}$ where

$$\eta_1 = \begin{bmatrix} -1 \\ 0 \\ 1 \end{bmatrix}, \eta_2 = \begin{bmatrix} 0 \\ 1 \\ 1 \end{bmatrix}, \eta_3 = \begin{bmatrix} 3 \\ -1 \\ 0 \end{bmatrix}.$$

Find the representation matrix $[T]_S$, where $S = \{\epsilon_1, \epsilon_2, \epsilon_3\}$ is the standard basis of $\mathbb{R}^3$.

2. Let $S = \{v_1, v_2, v_3, v_4\}$ be a basis of a vector space V. Let $T : V \to V$ be a linear transformation with

$$[T]_S = \begin{bmatrix} 1 & 3 & 1 & 0 \\ 0 & 2 & -6 & 0 \\ 0 & 0 & 4 & 0 \\ 0 & -9 & -3 & 1 \end{bmatrix}.$$

i) Find $[T]_N$ if N is a basis of V with the transition matrix from N to S given by

$$P = \begin{bmatrix} 1 & 3 & 1 & 0 \\ 0 & 2 & 3 & 0 \\ 0 & 0 & 3 & 0 \\ 0 & 0 & 0 & 1 \end{bmatrix}.$$

ii) Find $\ker(T)$ and $\mathrm{Range}(T)$.

iii) Find a basis of $\ker(T)$ and extend it into a basis of V.

iv) Find a basis of $\mathrm{Range}(T)$ and extend it into a basis of V.

3. Let $T : V \to V$ be a linear operator. Show that $TV \subset \ker(T)$ if and only of $T^2 = 0$.

4. Let V be a finite dimensional vector space. Let $T_1, T_2 : V \to V$ be linear operators. Show that $T_1 V \subset T_2 V$ if and only if there exists linear operator T on V such that $T_1 = T_2 T$.

5. Let V be a finite dimensional vector space. Let $T_1, T_2 : V \to V$ be linear operators. Show that $\ker T_1 \subset \ker T_2$ if and only if there exists linear operator T on V such that $T_2 = T_1 T$.

8.3   Invariant subspaces

Definition 8.3.1. Let V be a vector space and $W \subset V$ a subspace. Let $T : V \to V$ be a linear operator. If $TW \subset W$, that is, for every $w \in W$, $Tw \in W$, then we call W a T-**invariant subspace** of V.

Example 8.3.2. Let $T : V \to V$ be a linear operator. Then the linear space V and $\{0\}$ are T-invariant. □

Lemma 8.3.3. Let $T : V \to V$ be a linear operator. Then $\ker(T)$ and $\mathrm{Range}(T)$ are T-invariant subspaces of V.

Proof. For every $w \in \ker(T)$ we have $Tw = 0 \in \ker(T)$ and hence $T(\ker(T)) \subset \ker(T)$. For every $y \in \text{Range}(T)$, there exists $v \in V$ such that $y = Tv \in V$ and $Ty = T(Tv) \in \text{Range}(T)$.   □

We are interested in one-dimensional T-invariant subspaces, say, $W = \text{span}(v)$ with $v \neq 0$. If W is T-invariant, then for every $w = \alpha v \in W$ with $\alpha \neq 0$ a scalar, $Tw = kv$ for some scalar k. That is,

$$T(\alpha v) = kv \implies T(v) = \frac{k}{\alpha} v.$$

Definition 8.3.4. Let V be a vector space over the scalar field K and $T : V \to V$ a linear operator. If $x \neq 0$ is such that

$$Tx = \lambda x,$$

for some scalar $\lambda \in K$, we call x an eigenvector of T associated with the eigenvalue λ.

If V is finite dimensional, there exists a matrix representation $[T]$ under a given basis. Then $T(v) = \frac{k}{\alpha} v$ can be rewritten

$$[T][v] = \frac{k}{\alpha}[v],$$

where $[v]$ is the coordinate vector of v. Therefore, if $W = \text{span}(v)$ is a one-dimensional T-invariant subspace of V, then the spanning vector is an eigenvector of the representation matrix.

Conversely, if $v \neq 0$ is an eigenvector of T associated with the eigenvalue λ, then for every $w \in W = \text{span}(v)$, we have $w = \alpha v$ for some $\alpha \in K$ and that $Tw = \lambda w = \lambda \alpha v \in W$. That is, W is T-invariant.

We have arrived at

Lemma 8.3.5. Let $T : V \to V$ be a linear operator. W is a one-dimensional T-invariant subspace of V if and only if W is an eigenspace of T.

Remark 8.3.6. For linear operators on general vector spaces, we usually specify the scalar field on which we have eigenvalues and eigenvectors. For example, the map $T : \mathbb{R}^2 \to \mathbb{R}^2$ defined by

$$T(x, y) = \begin{bmatrix} 0 & -1 \\ 1 & 0 \end{bmatrix} \begin{bmatrix} x \\ y \end{bmatrix}$$

has no eigenvalues in the underlying scalar field $\mathbb{R}$. But T has eigenvalues $\lambda_{1,2} = \pm i$ in $\mathbb{C}$ if T is defined on $\mathbb{C}^2$.   □

We know that every linear operator T on a finite dimensional space V has a square matrix representation $[T]$, which is dependent on the choice of the basis of the vector space. A question is how to choose a proper basis of V such that the matrix representation $[T]$ is diagonal?

Suppose $S = [v_1 : v_2 : \cdots : v_n]$ is a basis of V. We seek for a basis $L = [w_1 : w_2 : \cdots : w_n]$ of V such that $[T]_L$ is diagonal. By Theorem 8.2.7, we have

$$[T]_L = P^{-1}[T]_S P,$$

where P is the transition matrix from L to S. To have a diagonal $[T]_L$ we need the columns of the invertible matrix P to be eigenvectors of $[T]_S$. Namely, we need only to choose $L = [w_1 : w_2 : \cdots : w_n]$ such that

$$[w_1 : w_2 : \cdots : w_n] = [v_1 : v_2 : \cdots : v_n]P.$$

In summary, we have

Lemma 8.3.7. Let $T : V \to V$ be a linear operator on an n dimensional vector space V with a basis $S = [v_1 : v_2 : \cdots : v_n]$. If $[T]_S$ is diagonalizable, then $N = [v_1 : v_2 : \cdots : v_n]P$ is a basis of V such that $[T]_N$ is diagonal, where P is such that

$$[T]_N = P^{-1}[T]_S P.$$

Example 8.3.8. Let $S = \{v_1, v_2, v_3, v_4\}$ be a basis of a vector space V. Let $T : V \to V$ be a linear transformation with

$$[T]_S = \begin{bmatrix} 1 & 3 & 1 & 0 \\ 0 & -1 & -1 & 0 \\ 0 & 0 & 4 & 0 \\ 0 & -1 & -3 & -2 \end{bmatrix}$$

i) Find $[T]_N$ if N is a basis of V with the transition matrix from N to S given by

$$P = \begin{bmatrix} 1 & 3 & 1 & 1 \\ 0 & 2 & 3 & 2 \\ 0 & 0 & 3 & 0 \\ 0 & 0 & 0 & 1 \end{bmatrix}.$$

ii) Find a basis L of V such that $[T]_L$ is diagonal.

Solution: i) Since P is the transition matrix from N to S, we have

$$\begin{aligned} [T]_N &= P^{-1}[T]_S P \\ &= \begin{bmatrix} 1 & 3 & 1 & 1 \\ 0 & 2 & 3 & 2 \\ 0 & 0 & 3 & 0 \\ 0 & 0 & 0 & 1 \end{bmatrix}^{-1} \begin{bmatrix} 1 & 3 & 1 & 0 \\ 0 & 27 & -6 & 0 \\ 0 & 0 & 4 & 0 \\ 0 & -9 & -3 & 18 \end{bmatrix} \begin{bmatrix} 1 & 3 & 1 & 1 \\ 0 & 2 & 3 & 2 \\ 0 & 0 & 3 & 0 \\ 0 & 0 & 0 & 1 \end{bmatrix} \end{aligned}$$

$$= \begin{bmatrix} 1 & 8 & 12 & 2 \\ 0 & 1 & 3 & 3 \\ 0 & 0 & 4 & 0 \\ 0 & -2 & -12 & -4 \end{bmatrix}.$$

ii) To find a basis L such that $[T]_L$ is diagonal, we need to find the eigenvalues of $[T]_S$ and a diagonalizing matrix Q whose columns are eigenvectors of $[T]_S$. Solving $\det([T]_S - \lambda I) = 0$, we have

$$(\lambda - 4)(\lambda + 2)(\lambda + 1)(\lambda - 1) = 0 \implies \lambda_1 = 4, \lambda_2 = 2, \lambda_3 = -1, \lambda_4 = 1.$$

Corresponding to the eigenvectors λ_i, $i = 1, 2, 3, 4$, the eigenvectors q_i are chosen:

$$q_1 = \begin{bmatrix} -2 \\ 3 \\ -15 \\ 7 \end{bmatrix}, \ q_2 = \begin{bmatrix} 0 \\ 0 \\ 0 \\ 1 \end{bmatrix}, \ q_3 = \begin{bmatrix} 3 \\ -2 \\ 0 \\ 2 \end{bmatrix}, \ q_4 = \begin{bmatrix} 1 \\ 0 \\ 0 \\ 0 \end{bmatrix}.$$

That is, if we choose $Q = [q_1 : q_2 : q_3 : q_4]$ as the transition matrix from basis L to S, where

$$L = [w_1 : w_2 : w_3 : w_4]$$

$$= [v_1 : v_2 : v_3 : v_n] \begin{bmatrix} -2 & 0 & 3 & 1 \\ 3 & 0 & -2 & 0 \\ -15 & 0 & 0 & 0 \\ 7 & 1 & 2 & 0 \end{bmatrix},$$

then $[T]_L = \begin{bmatrix} 4 & 0 & 0 & 0 \\ 0 & -2 & 0 & 0 \\ 0 & 0 & -1 & 0 \\ 0 & 0 & 0 & 1 \end{bmatrix}$, which is diagonal. $\qquad \square$

Theorem 8.3.9. Let $T : V \to V$ be a linear operator on an n dimensional vector space V over the complex numbers $\mathbb{C}$ with a basis $S = [v_1 : v_2 : \cdots : v_n]$. If $[T]_S$ is diagonalizable, then V assumes a direct sum decomposition of the eigenspaces:

$$V = W_{\lambda_1} \oplus W_{\lambda_2} \oplus \cdots \oplus W_{\lambda_n},$$

where W_{λ_i}, $i = 1, 2, \cdots, n$ are eigenspaces of T corresponding to the eigenvalues $\lambda_i, i = 1, 2, \cdots, n$.

Proof. Note that $[T]_S$ is diagonalizable and has n linearly independent eigenvectors. Then correspondingly T has n linearly independent eigenvectors which span the whole space V. That is, for every $x \in V$, there exists $x_i \in W_{\lambda_i}$ such that

$$x = x_1 + x_2 + \cdots + x_n.$$

Since the n eigenvectors are linearly independent, we have

$$W_{\lambda_i} \cap W_{\lambda_j} = \emptyset, \quad \text{if } i \neq j.$$

Therefore, we have

$$V = W_{\lambda_1} \oplus W_{\lambda_2} \oplus \cdots \oplus W_{\lambda_n}.$$

$\square$

Exercise 8.3.10.

1. Let $S = \{v_1,\, v_2,\, v_3\}$ be a basis of a vector space V. Let $T : V \to V$ be a linear transformation with

$$[T]_S = \begin{bmatrix} 1 & 3 & 1 \\ 0 & 27 & -6 \\ 0 & 0 & 4 \end{bmatrix}.$$

   i) Find $[T]_N$ if N is a basis of V with the transition matrix from N to S given by

$$P = \begin{bmatrix} 1 & 3 & 1 \\ 0 & 2 & 3 \\ 0 & 0 & 3 \end{bmatrix}.$$

   ii) Find a basis L of V such that $[T]_L$ is diagonal.

2. Let P_3 denote the vector space of all polynomials with real coefficients. Let $T : P_3 \to P_3$ be defined by

$$T(f)(x) = x^2 f''(x).$$

   i) Find $[T]$ under the standard basis $\{1, x, x^2, x^3\}$ of P_3;

   ii) Find all eigenvalues and eigenvectors of T.

3. Let $S, T : V \to V$ be linear operator on a vector space V. Suppose that $S \circ T = T \circ S$. Show that the range of S is a T-invariant subspace of V.

4. Let $T : V \to V$ be a linear operator on an n-dimensional vector space V over the real numbers $\mathbb{R}$ with a basis $S = [v_1 : v_2 : \cdots : v_n]$. If $[T]_S$ is symmetric, then V assumes a direct sum decomposition of the eigenspaces:

$$V = W_{\lambda_1} \oplus W_{\lambda_2} \oplus \cdots \oplus W_{\lambda_n},$$

where W_{λ_i}, $i = 1, 2, \cdots, n$ are eigenspaces of T corresponding to the eigenvalues $\lambda_i, i = 1, 2, \cdots, n$.

5. Let $T : V \to V$ be a linear operator on a vector space V. Let f, g be polynomials. Show that $f(T)g(T) = g(T)f(T)$.

6. Let $T : V \to V$ be a linear operators on a vector space V. Let f be a polynomial and $W = \ker f(T)$. Then W is a T-invariant subspace of V.

8.4 Decomposition of vector spaces

In the last section we have related the problem of diagonalization of the representation matrix of a linear operator on finite dimensional spaces to that of decomposition of a vector space into one-dimensional subspaces.

A natural question is what if the representation matrix is not diagonalizable? We show in the next two sections that in this case, the representation matrix can be reduced into **block diagonal form** and further to **Jordan normal form**, which is a block diagonal form of a matrix. Moreover, the vector space is decomposed into a direct sum of generalized eigenspaces associated with each of the Jordan blocks.

Theorem 8.4.1. Let p, p_1 and p_2 be polynomials which satisfy that

  i) $p = p_1 p_2$;

  ii) p_1 and p_2 are polynomials with degree larger than or equal to 1;

  iii) the greatest common divisor of p_1 and p_2 is 1.

Let $T : V \to V$ be a linear operator on a vector space V. If $p(T) = 0$, then
$$V = W_1 \oplus W_2,$$
where $W_1 = \ker(p_1(T))$ and $W_2 = \ker(p_2(T))$.

Proof. Since the greatest common divisor of p_1 and p_2 is 1, by the Euclidean algorithm, there exist polynomials q_1 and q_2 such that

$$1 = p_1(t)q_1(t) + p_2(t)q_2(t),$$

which lead to
$$I = p_1(T)q_1(T) + p_2(T)q_2(T).$$

Then for every $v \in V$, we have the following decomposition of v,

$$v = p_1(T)q_1(T)v + p_2(T)q_2(T)v, \tag{8.1}$$

where the first part satisfies

$$
\begin{aligned}
p_2(T)p_1(T)q_1(T)v &= p_1(T)p_2(T)q_1(T)v \\
&= q_1(T)p_1(T)p_2(T)v \\
&= q_1(T)p(T)v \\
&= 0,
\end{aligned}
$$

and the second part satisfies

$$p_1(T)p_2(T)q_2(T)v = q_2(T)p_1(T)p_2(T)v$$
$$= q_2(T)p(T)v$$
$$= 0.$$

That is, $p_1(T)q_1(T)v \in W_2$ and $p_1(T)q_1(T)v \in W_1$. To prove the uniqueness of the decomposition, let $v = w_1 + w_2 = w'_1 + w'_2$ with $w_1, w'_1 \in W_1$ and $w_2, w'_2 \in W_2$. Then we have

$$q_1(t)p_1(T)v = q_1(t)p_1(T)(w_1 + w_2), \quad q_1(t)p_1(T)v = q_1(t)p_1(T)(w'_1 + w'_2),$$

which lead to

$$q_1(t)p_1(T)v = q_1(t)p_1(T)w_2, \quad q_1(t)p_1(T)v = q_1(t)p_1(T)w'_2.$$

Applying (8.1) to w_2, w'_2 we have

$$w_2 = p_1(t)q_1(T)w_2, \quad w'_2 = p_1(t)q_1(T)w'_2.$$

Then we have

$$w_2 = q_1(t)p_1(T)v = w'_2,$$

and $w_1 = v - w_2 = v - w'_2 = w'_1$. That is, $w_1 = w'_1$, $w_2 = w'_2$. The decomposition is unique. $\qquad\square$

An example of the polynomial p such that $p(T) = 0$ can be seen from the Cayley-Hamilton theorem, where we know that if p is the characteristic polynomial of the representation matrix A of the linear operator T, then $p(A) = 0$. In the following we apply Theorem 8.4.1 to obtain a decomposition of a finite dimensional vector space according to the eigenspaces of a linear operator on the space.

Theorem 8.4.2. Let $T : V \to V$ be a linear operator on a finite dimensional vector space V over the scalar field of complex numbers. Assume that p is a polynomial with $p(T) = 0$ and that

$$p(t) = (t - \lambda_1)^{m_1}(t - \lambda_2)^{m_2} \cdots (t - \lambda_r)^{m_r},$$

is a factorization of p where $\lambda_i, i = 1, 2, \cdots, r$ are distinct roots of p. Let $W_i = \ker(T - \lambda_i I)^{m_i}$, $i = 1, 2, \cdots, r$. Then we have

$$V = W_1 \oplus W_2 \oplus \cdots \oplus W_r.$$

Proof. Let $p_i = (t - \lambda_1)^{m_1}(t - \lambda_2)^{m_2} \cdots (t - \lambda_i)^{m_i}$, $p_{-i} = (t - \lambda_{i+1})^{m_{i+1}}(t - \lambda_2)^{m_2} \cdots (t - \lambda_r)^{m_r}$ with $i = 1, 2, \cdots, r$. By Theorem 8.4.1, we have the following decomposition of V:

$$V = W_1 \oplus W_{-1},$$

where $W_{-1} = \ker(p_{-i}(T))$ is a subspace of V. Notice that $p_{-1}(T)w = 0$ for every $w \in W_{-1}$. Regarding W_{-1} as the whole space, by Theorem 8.4.2, we have the following decomposition of W_{-1}:

$$W_{-1} = W_2 \oplus W_{-2}.$$

It is known that W_2 is a subspace of V. W_{-2} is a subspace of W_{-1}; hence it is also a subspace of V. Then we have

$$V = W_1 \oplus W_2 \oplus W_{-2}.$$

By mathematical induction, we can then obtain that

$$V = W_1 \oplus W_2 \oplus \cdots \oplus W_r.$$

$\square$

Remark 8.4.3. By Theorem 8.4.2, we can group the bases of each of the subspaces W_i, $i = 1, 2, \cdots$ to obtain a basis of V. Then the representation matrix of T on V is a block diagonal form:

$$[T]_V = \begin{bmatrix} A_1 & 0 & \cdots & 0 \\ 0 & A_2 & \cdots & 0 \\ \vdots & \vdots & \ddots & \vdots \\ 0 & \cdots & \cdots & A_r \end{bmatrix},$$

where A_i is the representation matrix of T restricted to the subspaces $W_i = \ker(T - \lambda_i)^{m_i}$ which is called a **generalized eigenspace** of T. $\square$

Example 8.4.4. Let $T : V \to V$ be a linear operator on an n dimensional vector space. Suppose that $T^2 = I$. Show that

i) T has eigenvalues $\lambda_1 = 1$, $\lambda_2 = -1$.

ii) $V = W_{\lambda_1} \oplus W_{\lambda_2}$, where W_{λ_i}, $i = 1, 2$ are eigenspaces of T corresponding to the eigenvalues λ_i.

Solution: i) Let λ be a eigenvalue with v a corresponding eigenvector. Then we have

$$Tv = \lambda v \implies T^2v = \lambda Tv = \lambda^2 v.$$

Since $T^2 = T$, we have $T^2v = Tv = \lambda v$. Therefore we obtain

$$\lambda^2 v = \lambda v \implies \lambda = \pm 1.$$

ii) Let $p(t) = (t-1)(t+1)$. We have $p(T) = 0$. By Theorem 8.4.2, we have

$$V = W_{\lambda_1} \oplus W_{\lambda_2}$$

where W_{λ_i}, $i = 1, 2$ are eigenspaces of T corresponding to the eigenvalues λ_i. $\square$

Exercise 8.4.5.

1. Let $T : V \to V$ be a linear operator over an n dimensional vector space. Suppose that $T^2 = T$. Show that

    i) T has eigenvalues $\lambda_1 = 1$, $\lambda_2 = 0$.

    ii) $V = W_{\lambda_1} \oplus W_{\lambda_2}$, where W_{λ_i}, $i = 1$, 2 are eigenspaces of T corresponding to the eigenvalues λ_i.

2. Let $T : V \to V$ be a linear operator over an n dimensional vector space. Let λ_i, $i = 1, 2, \cdots, m$ be the set of all distinct eigenvalues of T. Suppose that the representation matrix $[T]$ under a basis of V is diagonalizable. Show that there exist linear operators T_i, $i = 1, 2, \cdots, m$ such that

    i) $T = \sum_{i=1}^{m} \lambda_i T_i$;

    ii) $I = \sum_{i=1}^{m} T_i$;

    iii) $T_i T_j = 0$ if $i \neq j$, $i, j \in \{1, 2, \cdots, m\}$;

    iv) $T_i^2 = T_i$, $i = 1, 2, \cdots, m$;

    v) $T_i V = V_{\lambda_i}$.

3. Let p, p_1 and p_2 be polynomials which satisfy that

    i) $p = p_1 p_2$;

    ii) p_1 and p_2 are polynomials with degree larger than or equal to 1;

    iii) the greatest common divisor of p_1 and p_2 is 1.

Let $T : V \to V$ be a linear operator on a vector space V. If $p(T) = 0$, then $\ker(p_1(T)) = \mathrm{Range}(p_2(T))$.

8.5   Jordan normal form

    We improve the results of the previous section to show that the blocks of the block diagonal decomposition of the representation matrix $[T]$ can be reduced to a type of triangular matrices called **Jordan blocks**, if the bases of the decomposed subspaces are properly chosen.

    Let $T : V \to V$ be a linear operator on the n dimensional vector space V and $v \in V$ be a nonzero vector. If we repeatedly use T to act on v to produce the sequence

$$v, Tv, T^2 v, \cdots,$$

then there exists a positive integer $k \in \mathbb{N}$ such that $T^k v = 0$, because of the finite dimensionality of V.

With the decomposition in Theorem 8.4.2, we seek basis of $W = \ker(T - \lambda I)^k$ where λ is an eigenvalue of T. Let $v \in W$ be such that $(T - \lambda I)^{k-1}v \neq 0$. Then $(T - \lambda I)^i v \neq 0$ for every $0 \leq i \leq k - 1$. Consider the vector equation

$$c_0 v + c_1(T - \lambda I)v + c_2(T - \lambda I)^2 v + \cdots + c_{k-1}(T - \lambda I)^{k-1}v = 0.$$

Applying $(T - \alpha I)^{k-1}$ repeatedly on both sides of the above vector equation and using the fact that v is k-periodic, we successively obtain $c_0 = 0$, $c_1 = 0, \cdots, c_{k-1} = 0$. Therefore, the set of vectors

$$\left\{ (T - \lambda I)^{k-1}v, (T - \lambda I)^{k-2}v, \cdots, (T - \lambda I)v, v \right\}$$

is linearly independent. We have shown

Lemma 8.5.1. Let $v \in V$ be such that $(T - \lambda I)^k v = 0$ but $(T - \lambda I)^{k-1}v \neq 0$. Then the set of vectors

$$\left\{ (T - \lambda I)^{k-1}v, (T - \lambda I)^{k-2}v, \cdots, (T - \lambda I)v, v \right\}$$

is linearly independent.

Lemma 8.5.1 implies that

$$\left\{ (T - \lambda I)^{k-1}v, (T - \lambda I)^{k-2}v, \cdots, (T - \lambda I)v, v \right\}$$

is an ordered basis for $W = \ker(T - \lambda I)^k$, if $v \in W$ is such that $(T - \lambda I)^{k-1}v \neq 0$. Note that

$$T(T - \lambda I)^i v = (T - \lambda I)^{i+1}v + \lambda(T - \lambda I)^i v, \text{ for every } 0 \leq i \leq k - 2,$$
$$T(T - \lambda I)^{k-1}v = (T - \lambda I)^k v + \lambda(T - \lambda I)^{k-1}v = \lambda(T - \lambda I)^{k-1}v.$$

Then the representation matrix of T on W with respect to this basis is

$$J = \begin{bmatrix} \lambda & 1 & 0 & \cdots & 0 \\ 0 & \lambda & 1 & \ddots & \vdots \\ 0 & \ddots & \ddots & \ddots & 0 \\ \vdots & \cdots & 0 & \lambda & 1 \\ 0 & \cdots & 0 & 0 & \lambda \end{bmatrix},$$

which is called the **Jordan block** associated with the eigenvalue λ of T. The basis

$$\left\{ (T - \lambda I)^{k-1}v, (T - \lambda I)^{k-2}v, \cdots, (T - \lambda I)v, v \right\}$$

is called a **Jordan basis** and the finite sequence of vectors is called a **Jordan chain**. The vector $(T - \lambda I)^{k-1}v$ is an eigenvector of T and is called the initial vector of the Jordan chain. v is called the end of the Jordan chain.

Suppose that V has been decomposed into

$$V = W_1 \oplus W_2 \oplus \cdots \oplus W_r,$$

where $W_i = \ker(T - \lambda_i)^{m_i}$, $i = 1, 2, \cdots, r$. If we choose a Jordan basis for each of the W_i, then the union of the Jordan basis is a basis for V and the representation matrix of T on V is then

$$[T] = \begin{bmatrix} J_1 & 0 & \cdots & 0 \\ 0 & J_2 & \ddots & \vdots \\ \vdots & \ddots & \ddots & 0 \\ 0 & \cdots & 0 & J_r \end{bmatrix},$$

where J_i, $i = 1, 2, \cdots, r$ are Jordan blocks. We call this form of $[T]$ the **Jordan normal form** for T.

Theorem 8.5.2. Let $T : V \to V$ be a linear operator on a finite dimensional vector space V over the scalar field of complex numbers. Then there exists a Jordan basis of V with which the representation matrix $[T]$ is in the Jordan normal form.

Proof. Let T be restricted to a generalized eigenspace $W = \ker(T - \lambda_i I)^{m_i}$. Then the representation matrix of T has only one eigenvalue, say λ. Let m be the least integer such that $(T - \lambda I)^m = 0$, namely, for every $x \in W$, $(T - \lambda I)^m w = 0$.

We show that there exists a Jordan basis of W such that the representation matrix of $[T]$ is in the Jordan normal form. Then by Theorem 8.4.2, the union of the Jordan bases of every generalized eigenspaces is such that the representation matrix of $[T]$ restricted to W is in Jordan normal form.

We proceed with induction on the dimension of $W = \ker(T - \lambda I)^m$. If $\dim W = 1$, the statement of the theorem is trivially true. Assume that for $\dim W \le n - 1$ the statement of the theorem holds. We prove it holds for $\dim W = n$.

Let $B = T - \lambda I$ and $r = \dim \ker B$. Note that there exists $w \in W$ with $B^{m-1} w \ne 0$ and $B^{m-1} w \in \ker B$. Then the formula $\dim BW + \dim \ker B = n$ implies that BW is a proper subspace of W and $\dim BW \le n-1$. By induction assumption, BW can be decomposed into subspaces W_i, $i = 1, 2, \cdots, k$:

$$BW = W_1 \oplus W_2 \oplus \cdots \oplus W_k,$$

and for each W_i, $i = 1, 2, \cdots, k$, there exists a Jordan chain,

$$B^{l_i - 1} u_i, \cdots, B u_i, u_i,$$

and the union of which is a Jordan basis of BW with $n - r$ vectors.

Next we extend the union of the Jordan basis of BW into a basis of W. Note for every $u_i \in BW$, $i = 1, 2, \cdots, k$, there exists a nonzero $v_i \in W$, such that $Bv_i = u_i$. Then the Jordan basis of BW is extended into

$$B^{l_i}v_i, B^{l_i-1}v_i, \cdots, v_i, i = 1, 2, \cdots, k.$$

Notice that $B^{l_i}v_i = B^{l_i-1}u_i \in \ker B$, $i = 1, 2, \cdots, k$. We can extend the set $\{B^{l_i}v_i\}_{i=1}^{k}$ into a basis of $\ker B$ by adding more vectors, say, $w_1, w_2, \cdots, w_{r-k}$, where each of the $w_i's$ is a Jordan chain of length 1.

It remains to show that

$$\bigcup_{i=1}^{k}\{v_i, Bv_i, \cdots, B^{l_i}v_i\} \cup \{w_i\}_{i=1}^{r-k} \tag{8.2}$$

is linearly independent. Consider the vector equation

$$\sum_{i=1}^{k}\sum_{j=1}^{l_i} c_{ij}B^j v_i + \sum_{j=1}^{r-k} d_j w_j = 0, \tag{8.3}$$

where the $c_{ij}'s$ and $d_i's$ are constants to be determined. Applying B on both sides of the vector equation, we have

$$\sum_{i=1}^{k}\sum_{j=1}^{l_i} c_{ij}B^{j+1} v_i = 0 = \sum_{i=1}^{k}\sum_{j=1}^{l_i} c_{ij}B^j u_i.$$

Note that $B^{l_i}u_i = 0$ for every $i = 1, 2, \cdots, k$. Then we have

$$\sum_{i=1}^{k}\sum_{j=1}^{l_i-1} c_{ij}B^j u_i = 0,$$

which by induction assumption leads to $c_{ij} = 0$ for every $i = 1, 2, \cdots, k$, $j = 1, 2, \cdots, l_i - 1$. Then by (8.3) we have

$$\sum_{i=1}^{k} c_{il_i}B^{l_i} v_i + \sum_{j=1}^{r-k} d_j w_j = 0,$$

which leads to $c_{il_i} = 0$, for $i = 1, 2, \cdots, k$ and $d_j = 0$, $j = 1, 2, \cdots, r - k$ since this is a linear combination of the basis of $\ker B$. Therefore, the set of n vectors at (8.2) is linearly independent and is a Jordan basis of W.

$\square$

Remark 8.5.3. If we identify a linear operator with its matrix representation, Theorem 8.5.2 indicates that every $n \times n$ matrix over the complex numbers has a Jordan normal form.

$\square$

Example 8.5.4. Let $A = \begin{bmatrix} -1 & -2 & 6 \\ -1 & 0 & 3 \\ -1 & -1 & 4 \end{bmatrix}$ be the representation matrix of a

linear operator $T : V \to V$. Find the Jordan normal form of A.

Solution: Let $p(t) = \det(A - \lambda I)$ be the characteristic polynomial. Solving $p(\lambda) = \det(A - \lambda I) = 0$, we have $-(\lambda - 1)^3 = 0$ and the eigenvalues $\lambda_{123} = 1$.

Solving $(A - \lambda I)x = 0$ we obtain two linearly independent eigenvectors:

$$u_1 = \begin{bmatrix} -1 \\ 1 \\ 0 \end{bmatrix}, u_2 = \begin{bmatrix} 3 \\ 1 \\ 0 \end{bmatrix}.$$

Since the algebraic multiplicity of the eigenvalue $\lambda = 1$ is 3, but geometrical multiplicity is 2, by Theorem 6.2.5 A is not diagonalizable. By Theorem 8.5.2, there exists a Jordan basis such that the representation matrix of T is in Jordan form, or, equivalently, A is similar to a matrix in Jordan form.

Next we compute the least number m such that $(A - \lambda I)^m = 0$. We have

$$A - I = \begin{bmatrix} -2 & -2 & 6 \\ -1 & -1 & 3 \\ -1 & -1 & 3 \end{bmatrix}, (A - I)^2 = 0.$$

This means that a basis of generalized eigenvectors consists of a Jordan chain of length 2 and one of length 1. The Jordan form is

$$J = \begin{bmatrix} 1 & 0 & 0 \\ 0 & 1 & 1 \\ 0 & 0 & 1 \end{bmatrix}$$

□

Exercise 8.5.5.

1. Let A be an $n \times n$ complex matrix. Show that A can be decomposed into the sum of a diagonal matrix and a nilpotent matrix. That is $A = D + N$, where D is diagonal and $N^m = 0$ for some $m \in \mathbb{N}$.

2. Let $A = \begin{bmatrix} 1 & 1 \\ \epsilon & 1 \end{bmatrix}$, where $\epsilon \in \mathbb{R}$ is a parameter. Show that if $\epsilon = 0$, A is not diagonalizable; otherwise, A is diagonalizable.

8.6 Computation of Jordan normal form

In the last section we have proved existence of Jordan normal form. We show in this section that the Jordan normal form is unique up to the order of Jordan blocks. We also develop a method to compute Jordan normal form.

Theorem 8.6.1. Let $T : V \to V$ be a linear operator on an n dimensional vector space V over the scalar field of complex numbers. Let $\lambda \in \mathbb{C}$ be an eigenvalue of T. Then for every positive integer $m \in \mathbb{N}$, the number of $m \times m$ Jordan blocks

$$J_{\lambda, m} = \begin{bmatrix} \lambda & 1 & 0 & \cdots & 0 \\ 0 & \lambda & 1 & \ddots & \vdots \\ \vdots & \ddots & \ddots & \ddots & 0 \\ 0 & \cdots & 0 & \lambda & 1 \\ 0 & \cdots & 0 & 0 & \lambda \end{bmatrix}$$

is

$$N_m = \operatorname{rank}(T - \lambda I)^{m-1} - 2\operatorname{rank}(T - \lambda I)^m + \operatorname{rank}(T - \lambda I)^{m+1}.$$

Proof. Suppose that under a Jordan basis, T is in a Jordan normal form with Jordan blocks J_{λ_i, m_i}, $i = 1, 2, \cdots, k$ and $m_1 + m_2 + \cdots + m_k = n$,

$$[T] = \begin{bmatrix} J_{\lambda_1, m_1} & 0 & \cdots & 0 \\ 0 & J_{\lambda_2, m_2} & \ddots & \vdots \\ \vdots & \ddots & \ddots & 0 \\ 0 & \cdots & 0 & J_{\lambda_k, m_k} \end{bmatrix},$$

where the Jordan blocks associated with the same eigenvalue are grouped together. Then the map $T - \lambda I$, where λ is one of the eigenvalues, has a Jordan normal form with two types of Jordan blocks. One type is those with zero main diagonals, and the other with nonzero main diagonals. Namely, we have

$$[T - \lambda I] = \begin{bmatrix} J_{0, m_1} & 0 & \cdots & & \cdots & & 0 \\ 0 & \ddots & 0 & & & & \vdots \\ \vdots & \ddots & J_{0, m_i} & \ddots & & & \vdots \\ \vdots & & \ddots & J_{\lambda_{i+1}, m_{i+1}} & \ddots & & 0 \\ \vdots & & & & \ddots & \ddots & 0 \\ 0 & \cdots & \cdots & & \cdots & 0 & J_{\lambda_k, m_k} \end{bmatrix},$$

where we placed the blocks with zero main diagonals in the left uppermost positions just for convenience of visualization.

Notice that $\operatorname{rank}(J_{0, m_i}^j) = m_i - j$ for every $0 \le j \le m_i$ and $J_{0, m_i}^{m_i} = 0$. Moreover, for every Jordan block J_{μ, m_i} in $[T - \lambda I]$ with $\mu \ne 0$, $\operatorname{rank}(J_{\mu, m_i}^j) = m_i$, for every positive $j \in \mathbb{N}$.

For every $m \in \mathbb{N}$, we note that $\text{rank}(J_{0,m}^m) = 0$, $\text{rank}(J_{0,m+1}^m) = 1$, $\text{rank}(J_{0,m+2}^m) = 2$ and so on. Then we have

$$\text{rank}(T - \lambda I)^m = N_{m+1} + 2N_{m+2} + \cdots + (n-m)N_n + \sum_{\lambda_k \neq 0} \text{rank}(J_{\lambda_k, m_k}^m).$$

By the same token, we have

$$\text{rank}(T - \lambda I)^{m+1} = N_{m+2} + 2N_{m+3} + \cdots + (n-m-1)N_n + \sum_{\lambda_k \neq 0} \text{rank}(J_{\lambda_k, m_k}^m).$$

Then we have

$$\text{rank}(T - \lambda I)^m - \text{rank}(T - \lambda I)^{m+1} = N_{m+1} + N_{m+2} + \cdots + N_n. \qquad (8.4)$$

Since m is arbitrary, we have

$$\text{rank}(T - \lambda I)^{m-1} - \text{rank}(T - \lambda I)^m = N_m + N_{m+1} + \cdots + N_n. \qquad (8.5)$$

Note that if $m = 1$, (8.5) is valid and we have $\text{rank}(T - \lambda I)^{m-1} = \text{rank}(T - \lambda I)^0 = n$. This is because each Jordan block with all zeros in the main diagonal is 1 rank in deficiency from the full rank and $N_m + N_{m+1} + \cdots + N_n$ counts the total rank deficiencies for every Jordan block. Therefore, (8.5) combined with (8.4) gives

$$N_m = \text{rank}(T - \lambda I)^{m-1} - 2\,\text{rank}(T - \lambda I)^m + \text{rank}(T - \lambda I)^{m+1}.$$

$\square$

Remark 8.6.2. By Theorem 8.6.1 we know that the number of a specific size of Jordan blocks is independent of the choice of the Jordan basis since representation matrices under different bases are similar. Hence the Jordan normal form is unique up to the order of the Jordan blocks. $\square$

Example 8.6.3. Let $A = \begin{bmatrix} -1 & -2 & 6 \\ -1 & 0 & 3 \\ -1 & -1 & 4 \end{bmatrix}$ be the representation matrix of a linear operator $T : V \to V$. Find the number of 1×1, 2×2 and 3×2 Jordan blocks of A.

Solution: Let $p(t) = \det(A - \lambda I)$ be the characteristic polynomial. Solving $p(\lambda) = \det(A - \lambda I) = 0$, we have $-(\lambda - 1)^3 = 0$ and the eigenvalues $\lambda_{123} = 1$. By Theorem 8.6.1, we know that for the eigenvalue $\lambda_{+123} = 1$, we have

$$N_1 = \text{rank}(A - I)^0 - 2\text{rank}(A - I)^1 + \text{rank}(A - I)^2,$$
$$N_2 = \text{rank}(A - I)^1 - 2\text{rank}(A - I)^2 + \text{rank}(A - I)^3,$$
$$N_3 = \text{rank}(A - I)^2 - 2\text{rank}(A - I)^3 + \text{rank}(A - I)^3.$$

We have

$$A - I = \begin{bmatrix} -2 & -2 & 6 \\ -1 & -1 & 3 \\ -1 & -1 & 3 \end{bmatrix}, (A - I)^2 = 0.$$

We have $\text{rank}(A - I)^0 = 3$, $\text{rank}(A - I)^1 = 1$, $\text{rank}(A - I)^m = 0$ for $m \geq 2$. Therefore, we have

$$N_1 = 3 - 2 \cdot 1 + 0 = 1,$$
$$N_2 = 1 - 2 \cdot 0 + 0 = 1,$$
$$N_3 = 0.$$

The Jordan form is then

$$J = \begin{bmatrix} 1 & 0 & 0 \\ 0 & 1 & 1 \\ 0 & 0 & 1 \end{bmatrix}.$$

$\square$

Theorem 8.6.4. Let $x_1, x_2, \cdots, x_k$ be linearly independent eigenvectors of T corresponding to the same eigenvalue λ. Suppose that for $i = 1, 2, \cdots, k$, there exists u_i such that

$$x_i = (T - \lambda I)^{m_i - 1} u_i \neq 0, \quad T x_i = (T - \lambda I)^{m_i} u_i = 0.$$

Let J_i be the Jordan chain of $\{(T - \lambda I)^{m_i - 1} u_i, (T - \lambda I)^{m_i - 2} u_i, \cdots, u_i\}$. Then $J = \bigcup_{i=1}^{k} J_i$ is linearly independent.

Proof. We first show that $J_i \cap J_j = \emptyset$ if $i \neq j$. Otherwise, there exists k, l with $1 \leq k \leq m_i$ and $1 \leq l \leq m_j$ such that

$$(T - \lambda I)^{m_i - k} u_i = (T - \lambda I)^{m_j - l} u_j. \tag{8.6}$$

Then applying $(T - \lambda I)^{k-1}$ and $(T - \lambda I)^{l-1}$ on both sides of (8.6), respectively, we have

$$x_i = (T - \lambda I)^{m_j - l + k - 1} u_j, \ x_j = (T - \lambda I)^{m_i - k + l - 1} u_i.$$

If $k = l$, we have $x_i = x_j$ which is a contradiction since x_i and x_j are linearly independent eigenvectors. If $k > l$, then $m_j - l + k - 1 \geq m_j$ and we have $x_i = 0$, which is also a contradiction. If $k < l$, then $m_i - k + l - 1 \geq m_i$ and we have $x_j = 0$, which is a contradiction, too.

Next, we show that $\bigcup_{i=1}^{k} J_i$ is linearly independent. Let $W = \text{span} \bigcup_{i=1}^{k} J_i$ and restrict $B = T - \lambda I$ to W. Then BW is an invariant subspace of W. We proceed with mathematical induction on the number of vectors in $\bigcup_{i=1}^{k} J_i$.

If the number of vectors in $\bigcup_{i=1}^{k} J_i$ is less than or equal to 1, then the statement of the theorem is trivially true. Assume that the statement of the theorem is true with the number of vectors in $\bigcup_{i=1}^{k} J_i$ less than $n - 1$. We consider the statement with the number of vectors in $\bigcup_{i=1}^{k} J_i$ equal to n. Then we have $\dim W \leq n$.

Next we find a basis of BW. Let $J' = \bigcup_{i=1}^{k} J_i \setminus \{u_i\}$. Then on the one hand, we have $\text{span}(J') \subset BW$. On the other hand, for every $v \in BW$, there exists $w \in W$ such that $v = Bw \in \text{span}(J')$. Therefore, $\text{span}(J') = BW$ and by induction, J' is a basis for BW since the initial vector of the Jordan chains is not changed from J to J'. It follows that $\dim BW = n - k$.

Note that the k initial vectors of the Jordan chains in J' are in $\ker B$ and we have $\dim \ker B \geq k$. By Theorem 8.2.3, we have

$$n \geq \dim W = \dim BW + \dim \ker B \geq n - k + k = n.$$

It follows that $\dim W = \dim \text{span}(J) = n$ and J is linearly independent. $\qquad\square$

The next question is how to find a Jordan basis to reduce a representation matrix into Jordan form. We may proceed with the following steps.

(1) Find the number of Jordan blocks of every possible size of m and determine the Jordan normal form.

(2) For every eigenvalue λ and every Jordan block size m, compute u_1 such that
$$(T - \lambda I)^m u_1 = 0, \quad (T - \lambda I)^{m-1} u_1 \neq 0.$$
Then we obtain a Jordan chain of
$$\{(T - \lambda I)^{m-1} u_1, \, (T - \lambda I)^{m-2} u_1, \cdots, u_1\}.$$
Notice that $(T - \lambda I)^{m-1} u_1$ is an eigenvector of T.

(3) Group all Jordan chains to form a Jordan basis, or equivalently an invertible matrix M such that $M^{-1}[T]M$ is in Jordan normal form.

Example 8.6.5. Let $A = \begin{bmatrix} -1 & -2 & 6 \\ -1 & 0 & 3 \\ -1 & -1 & 4 \end{bmatrix}$ be as in Example 8.6.3. From Example 8.6.3, we know that the eigenvalues are $\lambda_{123} = 1$ and the Jordan normal form is

$$J = \begin{bmatrix} 1 & 0 & 0 \\ 0 & 1 & 1 \\ 0 & 0 & 1 \end{bmatrix}.$$

For the Jordan block with size 1, we solve for an eigenvector with

$$(A - I)u = \begin{bmatrix} -2 & -2 & 6 \\ -1 & -1 & 3 \\ -1 & -1 & 3 \end{bmatrix} \begin{bmatrix} x \\ y \\ z \end{bmatrix} = 0,$$

which lead to

$$u_1 = \begin{bmatrix} -1 \\ 1 \\ 0 \end{bmatrix}, \ u_1' = \begin{bmatrix} 3 \\ 0 \\ 1 \end{bmatrix}.$$

Either $\{u_1\}$ or $\{u_1'\}$ forms a Jordan chain of length 1.

For the Jordan block with size 2, we have $(A - I) \neq 0$ and $(A - I)^2 = 0$. Take an arbitrary nonzero vector, say $u_2 = (1, 0, 0)$ such that $(A - I)u_2 \neq 0$ and set

$$u_2^1 = (A - I)u_2 = \begin{bmatrix} -2 \\ -1 \\ -1 \end{bmatrix}.$$

Then $\{u_2^1, u_2\}$ forms a Jordan chain of length 2.

According to Theorem 8.6.4, we need to choose linearly independent initial vectors for the Jordan chains. Indeed, by inspection we note that both $\{u_2^1, u_1\}$ and $\{u_2^1, u_1'\}$ are linearly independent. Therefore, we may choose

$$M = [u_1 : u_2^1 : u_2] = \begin{bmatrix} -1 & -2 & 1 \\ 1 & -1 & 0 \\ 0 & -1 & 0 \end{bmatrix},$$

or

$$M = [u_1' : u_2^1 : u_2] = \begin{bmatrix} 3 & -2 & 1 \\ 0 & -1 & 0 \\ 1 & -1 & 0 \end{bmatrix},$$

such that

$$M^{-1}AM = J.$$

Example 8.6.6. Find the Jordan normal form J of

$$A = \begin{bmatrix} 2 & 0 & 0 & 0 \\ 0 & 2 & 1 & 0 \\ 0 & 0 & 1 & 1 \\ 0 & 0 & 0 & 1 \end{bmatrix},$$

and find M such that $M^{-1}AM = J$.

Solution: The characteristic polynomial is $\det(A - \lambda I) = (\lambda - 2)^2(\lambda - 1)^2$ and eigenvalues are $\lambda_{12} = 2$ and $\lambda_{34} = 1$. For each eigenvalue, there are possibly 1×1 and 2×2 Jordan blocks. We use Theorem 8.6.1 to determine the number of each possible Jordan block.

For $\lambda_{12} = 2$, we consider $(A - 2I)$ and have

$$(A - 2I) = \begin{bmatrix} 0 & 0 & 0 & 0 \\ 0 & 0 & 1 & 0 \\ 0 & 0 & -1 & 1 \\ 0 & 0 & 0 & -1 \end{bmatrix}, \ (A - 2I)^2 = \begin{bmatrix} 0 & 0 & 0 & 0 \\ 0 & 0 & -1 & 1 \\ 0 & 0 & 1 & -2 \\ 0 & 0 & 0 & 1 \end{bmatrix},$$

and

$$(A - 2I)^3 = \begin{bmatrix} 0 & 0 & 0 & 0 \\ 0 & 0 & 1 & -2 \\ 0 & 0 & -1 & 3 \\ 0 & 0 & 0 & -1 \end{bmatrix}.$$

Then $\text{rank}(A - 2I) = \text{rank}(A - 2I)^2 = \text{rank}(A - 2I)^3 = 2$. By Theorem 8.6.1, for the eigenvalue $\lambda_{12} = 2$, we have

$$N_1 = \text{rank}(A - I)^0 - 2\text{rank}(A - I)^1 + \text{rank}(A - I)^2 = 4 - 4 + 2 = 2,$$
$$N_2 = \text{rank}(A - I)^1 - 2\text{rank}(A - I)^2 + \text{rank}(A - I)^3 = 2 - 4 + 2 = 0.$$

For $\lambda_{34} = 1$, we consider $(A - I)$ and have

$$(A-I) = \begin{bmatrix} 1 & 0 & 0 & 0 \\ 0 & 1 & 1 & 0 \\ 0 & 0 & 0 & 1 \\ 0 & 0 & 0 & 0 \end{bmatrix}, (A-I)^2 = \begin{bmatrix} 1 & 0 & 0 & 0 \\ 0 & 1 & 1 & 1 \\ 0 & 0 & 0 & 0 \\ 0 & 0 & 0 & 0 \end{bmatrix}, (A-I)^3 = \begin{bmatrix} 1 & 0 & 0 & 0 \\ 0 & 1 & 1 & 1 \\ 0 & 0 & 0 & 0 \\ 0 & 0 & 0 & 0 \end{bmatrix}.$$

Then $\text{rank}(A - I) = 3, \text{rank}(A - I)^2 = \text{rank}(A - I)^3 = 2$. By Theorem 8.6.1, for the eigenvalue $\lambda_{12} = 2$, we have

$$N_1 = \text{rank}(A - I)^0 - 2\text{rank}(A - I)^1 + \text{rank}(A - I)^2 = 4 - 6 + 2 = 0,$$
$$N_2 = \text{rank}(A - I)^1 - 2\text{rank}(A - I)^2 + \text{rank}(A - I)^3 = 3 - 4 + 2 = 1.$$

Therefore, the Jordan normal form is

$$J = \begin{bmatrix} 2 & 0 & 0 & 0 \\ 0 & 2 & 0 & 0 \\ 0 & 0 & 1 & 1 \\ 0 & 0 & 0 & 1 \end{bmatrix}.$$

Next we find M such that $M^{-1}AM = J$. For $\lambda_{12} = 2$, we solve $(A - 2I)x = 0$ to obtain the corresponding eigenspace

$$N(A - 2I) = \text{span}(u_1, u_1') = \text{span} \left\{ \begin{bmatrix} 1 \\ 0 \\ 0 \\ 0 \end{bmatrix}, \begin{bmatrix} 0 \\ 1 \\ 0 \\ 0 \end{bmatrix} \right\},$$

where $\{u_1\}$ and $\{u_1'\}$ are Jordan chains of length 1.

For $\lambda_{34} = 1$, we solve $(A - I)^2 x = 0$ and $(A - I)x \neq 0$ to obtain

$$u_2 = (0, -1, 1, 0), \ u_2' = (0, -1, 0, 1).$$

If we choose u_2 to form a Jordan chain of length 2, then $(A-I)u_2 = (0, 0, 0, 0)$ which is not desired. If we choose u_2' to form a Jordan chain of length 2, then

$u_2^1 = (A - I)u_2' = (0, -1, 1, 0) \neq 0$ and we obtain the desired Jordan chain $\{u_2^1, u_2'\}$ of length two. Then $\{u_1, u_1', u_2^1, u_2'\}$ is a Jordan basis for A. That is,

$$M = \begin{bmatrix} 1 & 0 & 0 & 0 \\ 0 & 1 & -1 & -1 \\ 0 & 0 & 1 & 0 \\ 0 & 0 & 0 & 1 \end{bmatrix}$$

is such that $M^{-1}AM = J$.

$\square$

Exercise 8.6.7.

1. Find the Jordan normal form of the following matrix.

$$\begin{bmatrix} 2 & 2 \\ 0 & 2 \end{bmatrix}, \quad \begin{bmatrix} 3 & 0 \\ 3 & 3 \end{bmatrix}, \quad \begin{bmatrix} 4 & 4 \\ 4 & 4 \end{bmatrix} \quad \begin{bmatrix} 5 & -5 \\ 5 & -5 \end{bmatrix}.$$

2. For each of the following matrices A, find a matrix M such that $M^{-1}AM$ is in Jordan normal form.

$$\begin{bmatrix} 1 & 0 & 0 & 0 \\ 0 & 1 & -1 & -1 \\ 0 & 0 & 1 & 0 \\ 0 & 0 & 0 & 1 \end{bmatrix}, \quad \begin{bmatrix} 1 & 1 & -1 \\ -2 & -2 & 2 \\ -3 & -3 & 3 \end{bmatrix}.$$

3. Use the Jordan normal form to prove the Cayley-Hamilton Theorem.

4. Let A be a square matrix with all eigenvalues $\{\lambda_1, \lambda_2, \cdots, \lambda_k\}$. Show that for every positive $m \in \mathbb{N}$, the eigenvalues of A^m are $\{\lambda_1^m, \lambda_2^m, \cdots, \lambda_k^m\}$.

5. Let A be a square matrix. Show that the eigenvalues of A are all zeros if and only if there exists a positive $m \in \mathbb{N}$ such that $A^m = 0$.

6. Let A be a square matrix. Show that if there exists a positive $m \in \mathbb{N}$ such that $A^m = 0$, then $\det(A + E) = 1$.

7. Let A be a square matrix with $A^m = I$ for some positive $m \in \mathbb{N}$. Show that A is diagonalizable and every eigenvalue of A is a root of $p(t) = t^m - 1$.

8. Let A be a square matrix with $A^2 = A$. Show that A is diagonalizable and every eigenvalue of A is a root of $p(t) = t^2 - t$.

9. Let A be an $n \times n$ matrix with 0 the m times repeated eigenvalue. Show that $\text{rank}(A) = \text{rank}(A^2)$ if and only if $\text{rank}(A) = n - m$.

10. Let

$$J = \begin{bmatrix} \lambda & 1 & 0 & \cdots & 0 \\ 0 & \lambda & 1 & \ddots & \vdots \\ \vdots & \ddots & \ddots & \ddots & 0 \\ 0 & \cdots & 0 & \lambda & 1 \\ 0 & \cdots & 0 & 0 & \lambda \end{bmatrix}$$

be an $n \times n$ Jordan block, p a polynomial. Show that

$$p(J) = \begin{bmatrix} p(\lambda) & \frac{1}{1!}p'(\lambda) & \frac{1}{2!}p''(\lambda) & \cdots & \frac{1}{(n-1)!}p^{(n-1)}(\lambda) \\ 0 & p(\lambda) & \frac{1}{1!}p'(\lambda) & \ddots & \frac{1}{(n-2)!}p^{(n-2)}(\lambda) \\ \vdots & \ddots & \ddots & \ddots & 0 \\ 0 & \cdots & 0 & p(\lambda) & \frac{1}{1!}p'(\lambda) \\ 0 & \cdots & 0 & 0 & p(\lambda) \end{bmatrix}.$$

Chapter 9

Linear programming

Many optimization problems in management and industry are modeled in the form of optimizing a linear function over the solution set of a system of linear equations.

$$\begin{aligned} \operatorname*{minimize}_{x} \quad & d^T x \\ \text{subject to} \quad & Ax = b, \text{ and } x \geq 0, \end{aligned} \tag{9.1}$$

where d, $x \in \mathbb{R}^n$, A is an $m \times n$ real matrix, $b \in \mathbb{R}^m$ and $x \geq 0$ means that each coordinate of x is nonnegative. Certainly when $m = n$ and A is invertible, we may first solve for x and obtain a unique optimal solution if $x \geq 0$ is satisfied at the same time. The issue is that in practice we may have

$$\operatorname{rank}(A) = m < n, \tag{9.2}$$

and hence $Ax = b$ has infinitely many solutions. It becomes a nontrivial task to examine the nonnegativity and optimality among infinitely many solutions. We call the optimization problem at (9.1) a **linear programming** problem. We devote this chapter to a short illustration of the mathematical theories and techniques related to solving this optimization problem.

9.1   Extreme points

Let us begin with an example.

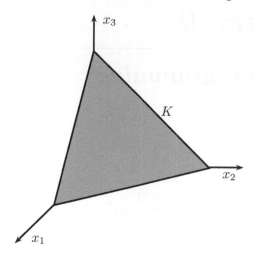

FIGURE 9.1: Feasible region for (9.20)

Example 9.1.1. Consider

$$\underset{x}{\text{minimize}} \quad x_1 + x_2$$

$$\text{subject to} \quad \begin{bmatrix} 1 & 1 & 1 \end{bmatrix} \begin{bmatrix} x_1 \\ x_2 \\ x_3 \end{bmatrix} = 1, \text{ and } x = (x_1, x_2, x_3) \geq 0. \tag{9.3}$$

The linear equation is the plane $x_1 + x_2 + x_3 = 1$ and is under the assumption that $x \geq 0$ gives a triangular region K in $\mathbb{R}^3$, as is shown in Figure 9.1. Every point in the triangular region K which satisfies the constraints is called a **feasible solution** of this linear programming problem. To find a minimizing point for the linear function $f(x) = x_1 + x_2$, which is called the **objective function**, we can do some elementary analysis from the constraints: $x_1 + x_2 \geq 0$ since $x \geq 0$ and the equality holds only if $x_1 = x_2 = 0$, $x_3 = 1$. Then we located the minimizing point $x^* = (0, 0, 1)$ such that the optimal value $f(x^*) = 0 + 0 = 0$.

Among infinitely many points in the region K we successfully find the optimal solution, which is located in a corner of the feasible region K. It is not by chance that the optimal solution is in the corner. Imagine that we hold the family of the planes $x_1 + x_2 = c$, $c \in \mathbb{R}$, and move it according to different values of c which becomes closer to the origin if the value of c becomes smaller. The value of c becomes minimal when the plane $x_1 + x_2 = c$ touches the corner $(0, 0, 1)$. It seems that the corner points are very special. Indeed, they are not a linear combination of any other vectors in the feasible region and are called **extreme points**. □

Definition 9.1.2. A set C in a vector space V is called a **convex set** if for every x_1, $x_2 \in C$ and for every $\lambda \in (0, 1)$, we have

$$\lambda x_1 + (1 - \lambda)x_2 \in C.$$

That is, the line segment between x_1 and x_2 is completely contained in C. We call $sy_1 + (1 - s)y_2$ with y_1, $y_2 \in V$ and $s \in (0, 1)$ a **convex combination** of y_1, y_2.

A point x in a convex set C is called an **extreme point** if x is not a convex combination of any two distinct vectors in C.

Example 9.1.3. Let A be an $m \times n$ real matrix with $\text{rank}(A) = m < n$ and $b \in \mathbb{R}^m$. Let

$$K = \{x \in \mathbb{R}^n : Ax = b, \text{ and } x \geq 0\}.$$

Then K is a convex set in $\mathbb{R}^n$. Indeed, for every x_1, $x_2 \in K$, we have $x_1 \geq 0$, $x_2 \geq 0$, and

$$Ax_1 = b, \quad Ax_2 = b.$$

It follows that for every $\lambda \in (0, 1)$, $\lambda x_1 + (1 - \lambda)x_2 \geq 0$ and

$$A(\lambda x_1 + (1 - \lambda)x_2) = \lambda Ax_1 + (1 - \lambda)Ax_2 = \lambda b + (1 - \lambda)b = b.$$

That is, x_1, $x_2 \in K$ and every $\lambda \in (0, 1)$, the convex combination $\lambda x_1 + (1 - \lambda)x_2 \in K$. K is a convex set in $\mathbb{R}^n$. $\square$

Next question is how to describe the extreme points of a feasible solution set defined by $K = \{x \in \mathbb{R}^n : Ax = b, \text{ and } x \geq 0\}$?

Theorem 9.1.4. Let A be an $m \times n$ real matrix with $\text{rank}(A) = m < n$ and $b \in \mathbb{R}^m$. Let

$$K = \{x \in \mathbb{R}^n : Ax = b, \text{ and } x \geq 0\}.$$

Then $x \in K$ is an extreme point of K if and only if $x = (x_1, x_2, \cdots, x_n) \in K$ has at most m nonzero coordinates and the columns of A corresponding to the nonzero coordinates of x for Ax are linearly independent.

Proof. "$\Longrightarrow$" Suppose that $x = (x_1, x_2, \cdots, x_n) \in K$ is an extreme point of K. Without loss of generality, assume that the first k coordinates of x are nonzero. Then we have

$$x_1 c_1 + x_2 c_2 + \cdots + x_k c_k = b,$$

where $c_1, c_2, \cdots, c_k$ are the corresponding columns of A for Ax with $A = [c_1 : c_2 : \cdots : c_n]$. We show that $c_1, c_2, \cdots, c_k$ are linearly independent and hence $k \leq m = \text{rank}(A)$.

Suppose not. Then there exists $(y_1, y_2, \cdots, \cdots, y_k) \neq 0$ such that

$$y_1 c_1 + y_2 c_2 + \cdots + y_k c_k = 0.$$

Let $y = (y_1, y_2, \cdots, \cdots, y_k, 0, \cdots, 0)$. Then there exists $\epsilon > 0$ small enough such that

$$x + \epsilon y \geq 0, \ x - \epsilon y \geq 0,$$

since $x \geq 0$ and $x_i > 0$ for $1 \leq i \leq k$. Moreover, we have $A(x + \epsilon y) = b = A(x - \epsilon y)$ which imply that $x + \epsilon y, x - \epsilon y \in K$ and

$$x = \frac{1}{2}(x + \epsilon y) + \frac{1}{2}(x - \epsilon y).$$

That is, x is a convex combination of two distinct points in K and x is not an extreme point in K. This is a contradiction.

"$\Longleftarrow$" Without loss of generality, let the first $k \leq m$ coordinates of x be nonzero with $x = (x_1, x_2, \cdots, x_k, 0, \cdots, 0)$ and the first k columns of $A = [c_1 : c_2 : \cdots : c_n]$ are linearly independent. Then we have

$$x_1 c_1 + x_2 c_2 + \cdots + x_k c_k = b. \tag{9.4}$$

Suppose, for contradiction, that x is not an extreme point. Then there exist $y, z \in K$ with $y \neq z$ and $0 < s < 1$ such that

$$x = sy + (1 - s)z.$$

Since $y \geq 0, z \geq 0$ and the last $n - k$ coordinates of x zero, it follows that the last $n - k$ coordinates of y and z are zero, too. Then we have $y = (y_1, y_2, \cdots, y_k, 0, \cdots, 0) \in K$, $z = (z_1, z_2, \cdots, z_k, 0, \cdots, 0) \in K$ and

$$y_1 c_1 + y_2 c_2 + \cdots + y_k c_k = b, \ z_1 c_1 + z_2 c_2 + \cdots + z_k c_k = b. \tag{9.5}$$

Since $\{c_1, c_2, \cdots, c_k\}$ are linearly independent, by (9.4) and (9.5) we have

$$x = y = z.$$

This is a contradiction. □

Definition 9.1.5. Let A be an $m \times n$ real matrix with $\text{rank}(A) = m < n$ and $b \in \mathbb{R}^m$. Let

$$K = \{x \in \mathbb{R}^n : Ax = b, \text{ and } x \geq 0\}.$$

Let B be an $m \times m$ invertible submatrix of A whose columns constitute a basis for the column space of A. If $x \in K$ assumes zero values for all $n - m$ coordinates not associated with B, x is called a **basic feasible solution** with respect to the basis B. The coordinates of basic feasible solution x associated with B are called **basic variables**. The coordinates of basic feasible solution x not associated with B are called **nonbasic variables**.

The conclusion of Theorem 9.1.4 implies that $x \in K$ is an extreme point if and only if x is a basic feasible solution with respect to some basis of the column space of A.

Now we discuss how a basic feasible solution is related to optimality.

Theorem 9.1.6. Let A be an $m \times n$ real matrix with rank$(A) = m < n$ and $b \in \mathbb{R}^m$. Let

$$K = \{x \in \mathbb{R}^n : Ax = b, \text{ and } x \geq 0\}.$$

i) If $K \neq \emptyset$, then there exists a basic feasible solution $x \in K$;

ii) If K contains an optimal solution, then it contains a basic feasible solution which is optimal.

Proof. i) Suppose that $K \neq \emptyset$. Then there exists $x = (x_1, x_2, \cdots, x_n) \in K$ such that

$$x_1 c_1 + x_2 c_2 + \cdots + x_n c_n = b,$$

where $A = [c_1 : c_2 : \cdots : c_n]$. Assume that x has k nonzero coordinates. Without loss of generality, we assume that

$$x_1 c_1 + x_2 c_2 + \cdots + x_k c_k = b. \tag{9.6}$$

If $\{c_1, c_2, \cdots, c_k\}$ is linearly independent, then $k \leq m$ and x is a basic feasible solution.

If $\{c_1, c_2, \cdots, c_k\}$ is linearly dependent, there exists a nontrivial linear combination of $\{c_1, c_2, \cdots, c_k\}$ such that

$$y_1 c_1 + y_2 c_2 + \cdots + y_k c_k = 0. \tag{9.7}$$

Then for every $\epsilon \in \mathbb{R}$, we obtain from (9.6) and (9.7) that

$$(x_1 - \epsilon y_1) c_1 + (x_2 - \epsilon y_2) c_2 + \cdots + (x_k - \epsilon y_k) c_k = b.$$

Let

$$\epsilon_0 = \min_{i \in \{1, 2, \cdots, k\}} \left\{ \frac{x_i}{y_i} : y_i > 0 \right\} > 0.$$

Then $x - \epsilon_0 y \in K$ where $y = (y_1, y_2, \cdots, y_k)$ has at most $k - 1$ positive coordinates. By the same token we repeat the same process on $x - \epsilon_0 y \in K$ to obtain a feasible solution $x - \epsilon' y \in K$ until the corresponding columns of A are linearly independent and hence $x - \epsilon' y$ is a basic feasible solution.

ii) Let $x = (x_1, x_2, \cdots, x_n) \in K$ be an optimal feasible solution. Assume that x has k nonzero coordinates. Without loss of generality, we assume that

$$x_1 c_1 + x_2 c_2 + \cdots + x_k c_k = b. \tag{9.8}$$

If the columns $\{c_1, c_2, \cdots, c_k\}$ of A are linearly independent, then $k \leq m$ and x is a basic feasible solution and is optimal.

If the columns $\{c_1, c_2, \cdots, c_k\}$ of A are linearly dependent, by the same procedure as for the proof of i), we can reduce x into a basic feasible solution. It remains to show that $x - \epsilon y$ is optimal for every ϵ. Note that $x - \epsilon y$ is a feasible solution for every ϵ and the value of the objective function at $x - \epsilon y$ is

$$c^T x - \epsilon c^T y.$$

If $c^T y \neq 0$, we can choose $\epsilon \neq 0$ so that $c^T x - \epsilon' c^T y \neq c^T x$ and $x - \epsilon y$ is another optimal feasible solution. This is a contradiction. Hence $c^T y = 0$ and $x - \epsilon y$ is an optimal feasible solution for every ϵ, which can be reduced into an optimal basic feasible solution. □

Theorems 9.1.4 and 9.1.6 imply that we need only to search among the extreme points, or, equivalently, the basic feasible solutions for optimal solutions, instead of searching among infinitely many feasible solutions. For small scale problems we may visualize the feasible region to locate the extreme points and find the optimal solution. In the next section, we introduce the simplex method for solving linear programming problems which includes an approach to transfer from one extreme point to another without geometrical visualization of the feasible region.

Exercise 9.1.7.

1. Determine whether the following problem has a feasible solution.

$$\underset{x}{\text{minimize}} \quad x_1 + x_2$$

$$\text{subject to} \quad \begin{bmatrix} 1 & 1 & 1 \\ -1 & 1 & -1 \end{bmatrix} \begin{bmatrix} x_1 \\ x_2 \\ x_3 \end{bmatrix} = \begin{bmatrix} 1 \\ 2 \end{bmatrix}, \text{ and } x = (x_1, x_2, x_3) \geq 0.$$

2. Draw a graph of the feasible region of the following optimization problem and find an optimal solution.

$$\underset{x}{\text{minimize}} \quad x_1 - x_2$$

$$\text{subject to} \quad \begin{bmatrix} 1 & 1 & 1 \end{bmatrix} \begin{bmatrix} x_1 \\ x_2 \\ x_3 \end{bmatrix} = 1, \text{ and } x = (x_1, x_2, x_3) \geq 0.$$

3. Show that every vector space is a convex set.

4. Let V be a real vector space. Let $A = \{x_1, x_2, \cdots, x_n\}$ be a subset of V and

$$\text{co}(A) = \left\{ \sum_{i=1}^{n} \lambda_i x_i : \lambda_1 + \lambda_2 + \cdots + \lambda_n = 1, \ \lambda_i \geq 0, \ i = 1, 2, \cdots, n \right\}.$$

Show that $\text{co}(A)$ is a convex set. (We call it the **convex hull** of A.)

5. Let A be an $m \times n$ real matrix with $\text{rank}(A) = m < n$ and $b \in \mathbb{R}^m$. Let

$$K = \{x \in \mathbb{R}^n : Ax = b, \text{ and } x \geq 0\}.$$

Show that there are finitely many extreme points in K.

6. Let A be an $m \times n$ real matrix, $b, y \in \mathbb{R}^m$ and $d, x \in \mathbb{R}^n$. Consider the following two optimization problems:

$$\begin{aligned}
\underset{x}{\text{minimize}} \quad & d^T x \\
\text{subject to} \quad & Ax \leq b, \text{ and } x \geq 0
\end{aligned} \tag{A}$$

and

$$\begin{aligned}
\underset{(x, y)}{\text{minimize}} \quad & d^T x \\
\text{subject to} \quad & [A : I] \begin{bmatrix} x \\ y \end{bmatrix} = Ax + y = b, \text{ and } x \geq 0, \ y \geq 0.
\end{aligned} \tag{B}$$

Show that x^* is an optimal solution of (A) if and only if $(x^*, 0)$ is an optimal solution of (B).

9.2 Simplex method

In the last section, we learned that optimal solutions of linear programming are among the extreme points, or, equivalently, among the basic feasible solutions. The **simplex method** for solving linear programming problems developed by George B. Dantzig in 1947 is a procedure which transfers from a basic feasible solution to another until an optimality condition is satisfied.

Initial basic feasible solution

The first question is how to find the initial basic feasible solution in order to begin the procedure of the simplex method. Let A be an $m \times n$ real matrix with $\text{rank}(A) = m < n$ and $b \in \mathbb{R}^m$ and $c, x \in \mathbb{R}^n$. Consider

$$\begin{aligned}
\underset{x}{\text{minimize}} \quad & d^T x \\
\text{subject to} \quad & Ax = b, \text{ and } x \geq 0.
\end{aligned} \tag{9.9}$$

We assume $b \geq 0$ since we may multiply the corresponding equation by minus one if there exists a negative coordinate of b. However, it is not obvious to identify a basic feasible solution for problem (9.9). So we consider the following auxiliary problem

$$\underset{(x,\,u)}{\text{minimize}} \quad \sum_{i=1}^{m} u_i \tag{9.10}$$

$$\text{subject to} \quad Ax + u = b, \text{ and } x \ge 0, \, u \ge 0.$$

The auxiliary problem (9.10) has a trivial basic feasible solution $(x, u) = (0, b)$ from which we may proceed with the simplex method to find an optimal solution if it exists.

If the minimum value of (9.10) is zero with $u = 0$, then it has a basic feasible solution $(x^*, 0)$ where x^* is a basic feasible solution of problem (9.9). If the minimum value of (9.10) is nonzero then $u \ne 0$ and (9.9) has no feasible solution because, otherwise, (9.10) should have achieved zero minimum.

Transfer of extreme points

In the following, we illustrate how to transfer from one basic feasible solution to another. For notational convenience, we assume that $A = [I_m : c_m : \cdots : c_n]$ and problem (9.9) has a basic feasible solution $(x_B, 0) \in \mathbb{R}^n$ where $x_B = (x_1, x_2, \cdots, x_m)$. (In practice the columns of I_m may appear in any column of A.) Then we have

$$x_1 c_1 + x_2 c_2 + \cdots + x_m c_m = b. \tag{9.11}$$

Noticing that $\{c_1, c_2, \cdots, c_m\}$ are linearly independent, each column c_p, $p > m$ can be written as a linear combination of the columns $\{c_1, c_2, \cdots, c_m\}$ which are a basis for the column space of A. Namely,

$$a_{1p} c_1 + a_{2p} c_2 + \cdots + a_{mp} c_m = c_p. \tag{9.12}$$

For every $\epsilon > 0$, multiplying (9.12) by ϵ and subtracting from (9.11) we have

$$(x_1 - \epsilon a_{1p})c_1 + (x_2 - \epsilon a_{2p})c_2 + \cdots + (x_m - \epsilon a_{mp})c_m + \epsilon c_p = b. \tag{9.13}$$

If $\epsilon = 0$, we have the old basic feasible solution. If $\epsilon > 0$ changes from zero, the coefficients of the linear combination in (9.13) are positive, until ϵ reaches the value:

$$\epsilon_0 = \min_i \left\{ \frac{x_i}{a_{ip}} : a_{ip} > 0 \right\}. \tag{9.14}$$

Let i_0 be the row index where ϵ_0 is achieved. Then the basis vector c_{i_0} at the i_0-th column is to be moved out. Since $a_{i_0 p} > 0$, $\{c_1, c_2, \cdots, c_m, c_p\} \setminus \{c_{i_0}\}$ is linearly independent and we obtained a new basic feasible solution:

$$(x_1 - \epsilon_0 a_{1p}, \cdots, x_{i_0-1} - \epsilon_0 a_{i_0-1,p}, 0, x_{i_0+1} - \epsilon a_{i_0+1p},$$

$$x_m - \epsilon_0 a_{mp}, 0, \cdots, 0, \epsilon, 0, \cdots, 0), \tag{9.15}$$

where the ϵ is the p-th coordinate. Let us use an example to show the above process.

Example 9.2.1. Consider the system $Ax = b$ with augmented matrix given by

c_1	c_2	c_3	c_4	c_5	b
1	0	0	-2	1	3
0	1	0	4	①	2
0	0	1	5	-3	1

where $x = (3, 2, 1, 0, 0)$ is a basic feasible solution with the corresponding columns $\{c_1, c_2, c_3\}$ linearly independent. Namely, x_1, x_2, x_3 are basic variables. If we want to bring c_5 into the basis for the column space, we choose

$$\epsilon_0 = \min_i \left\{ \frac{x_i}{a_{i5}} : a_{i5} > 0 \right\}$$

$$= \min \left\{ \frac{3}{1}, \frac{2}{1} \right\}$$

$$= 2,$$

which is achieved at $x_2 = 2$, $a_{25} = 1$. Namely $i_0 = 2$. c_2 will be removed and $\{c_1, c_3, c_5\}$ is the new basis. Using the expressions at (9.15), the new basic feasible solution is

$$(3 - 2 \cdot 1, 2 - 2 \cdot 1, 1 - 2 \cdot (-3), 0, 2) = (1, 0, 7, 0, 2).$$

If we use the pivot a_{25} in c_5 to reduce other entries in the column to zero, we obtain

c_1	c_2	c_3	c_4	c_5	b
1	-1	0	-6	0	1
0	1	0	4	①	2
0	3	1	17	0	7

The new basic variables are x_1, x_3, x_5 whose values are contained in the last column for b. □

Remark 9.2.2. From Example 9.2.1, we know that if the augmented matrix $[A : b]$ is in the reduced row echelon form, there is a basic feasible solution contained in the last column. During the process of transferring from one extreme point to another, if we reduce the basis vectors to have only one nonzero entry, then a basic feasible solution can be constructed according to the order of the basic variables. For instance, in the second tableau, $\{c_1, c_3, c_5\}$ are basis columns and the corresponding basic feasible solution is $(1, 0, 7, 0, 2)$.

Remark 9.2.3. If none of the a_{ip}'s in (9.14) is positive, then all the coefficients of the column vectors in (9.13) increase, as ϵ increases, while no new basic feasible solution can be identified. However, this means that there exists feasible solutions with arbitrarily large coefficients and the feasible region K is unbounded.

Remark 9.2.4. It may happen that the $\epsilon_0 = \frac{x_{i_0}}{a_{i_0 p}} = 0$ in (9.14), which implies that the new and old basic variables during the transfer are both zero, and the objective function will not change value. In such a case the process of carrying out the simplex method may enter into a cycle. We call this case a degenerate case. However, cycling is not common and can be avoided in the coding practice.

Optimality condition

After we learn how to transfer from one extreme point to another, we need to know when optimality has been achieved so that the process should stop.

Let A be an $m \times n$ real matrix with $\text{rank}(A) = m < n$ and $b \in \mathbb{R}^m$ and $d, x \in \mathbb{R}^n$. Consider problem (9.9). Suppose that $(x_B, 0)$ with $x_B \in \mathbb{R}^m$ is a basic feasible solution and that $A = [I_m : c_{m+1} : \cdots c_n]$ which is achievable using elementary row operations and/or renaming of the variables.

The value of the objective function at $(x_B, 0) = (b, 0)$ is

$$z_0 = d_B^T b,$$

where $d_B^T = [d_1, d_2, \cdots, d_m]$. To justify the current basic feasible solution $x = (x_B, 0)$ is optimal, we need to show that any other possible feasible solution will not lower the value of the objective function. Let $\{e_1, e_2, \cdots, e_m\}$ be the standard basis of $\mathbb{R}^m$. We have for every feasible solution $x = (x_1, x_2, \cdots, x_n)$ of $Ax = b$ with $A = [I_m : c_{m+1} : \cdots : c_n]$,

$$x_1 e_1 + x_2 e_2 + \cdots + x_m e_m = b - x_{m+1} c_{m+1} - x_{m+2} c_{m+2} - \cdots - x_n c_n. \tag{9.16}$$

Multiplying both sides of (9.16) with d_B^T, we have

$$\sum_{i=1}^m d_i x_i = d_1 x_1 + d_2 x_2 + \cdots + d_m x_m$$
$$= d_B^T b - x_{m+1} d_B^T c_{m+1} - x_{m+2} d_B^T c_{m+2} - \cdots - x_n d_B^T c_n \tag{9.17}$$
$$= z_0 - x_{m+1} d_B^T c_{m+1} - x_{m+2} d_B^T c_{m+2} - \cdots - x_n d_B^T c_n.$$

Then we have

$$d^T x = z_0 + (d_{m+1} - d_B^T c_{m+1})x_{m+1} + (d_{m+2} - d_B^T c_{m+2})x_{m+1}$$
$$+ \cdots + (d_n - d_B^T c_n)x_n. \tag{9.18}$$

Notice that any other feasible solution x satisfies $x \geq 0$. If $r_j = d_j - d_B^T c_j \geq 0$ for every $j \in \{1, 2, \cdots, n\}$, we have the value of the objective function:

$$z = d^T x \geq z_0.$$

Namely z_0 is the optimal value achieved at the current basic feasible solution $(x_B, 0)$. We have arrived at the following optimality condition theorem.

Theorem 9.2.5. Let $A = [c_1 : c_2 : \cdots : c_n]$ be an $m \times n$ real matrix with $\text{rank}(A) = m < n$, $b \in \mathbb{R}^m$, $d, x \in \mathbb{R}^n$ and $d = [d_1, d_2, \cdots, d_n]$. Consider the linear programming problem

$$\underset{x}{\text{minimize}} \quad d^T x$$

$$\text{subject to} \quad Ax = b, \text{ and } x \geq 0.$$

If $(x_B, 0) \in \mathbb{R}^n$ is a basic feasible solution, and $r_j = d_j - d_B^T c_j \geq 0$ for every $j \in \{1, 2, \cdots, n\}$, then $(x_B, 0)$ is an optimal solution.

Remark 9.2.6. Suppose that the initial basic variables are corresponding to the first m columns of A, which is the $m \times m$ identity matrix I_m; then for $j = 1, 2, \cdots, m$, we have

$$r_j = d_j - d_B^j c_j = d_j - d_j = 0.$$

Namely, r_j corresponding to basic variables are zero. $\qquad \square$

Exercise 9.2.7.

1. Let A be an $m \times n$ real matrix with $\text{rank}(A) = m < n$ and $b \in \mathbb{R}^m$ and $d, x \in \mathbb{R}^n$. Consider

$$\underset{x}{\text{minimize}} \quad d^T x$$

$$\text{subject to} \quad Ax = b, \text{ and } x \geq 0.$$

Let x be a basic feasible solution with respect to a basis B which is an $m \times m$ submatrix of A and each column of B contains only one nonzero entry which is 1. Suppose that the i_0-th column of A contained in B is replaced with the p-th column of A not contained in B in order to transfer from one basic feasible solution $x = a_0$ to another basic feasible solution, resulting in a new equivalent system $A'x = b'$ where the new basic feasible solution corresponds to a basis which contains only one nonzero entry 1. Show that

$$a'_{ij} = a_{ij} - \frac{a_{ip}}{a_{i_0p}} a_{i_0j}, \quad i \neq i_0,$$

$$a'_{i_0j} = \frac{a_{i_0j}}{a_{i_0p}}.$$

9.3   Simplex tableau

We have discussed the theory for the simplex method in the last section. We discuss in this section the technical details how to carry out the algorithm of the simplex method. Since the method for the transfer of extreme points of $Ax = b$, $x \geq 0$ was shown in the last section on augmented matrix, what is left is the details on how to check the optimality conditions at each step of the transfer.

The optimality condition is derived from the objective function $z = d^T x$, or explicitly

$$d_1x_1 + d_2x_2 + \cdots + d_nx_n - z = 0. \tag{9.19}$$

If we want to append this equation to the augmented matrix $[A : b]$, then we need a separate column to record the coefficients of z during elementary row operations. However, this is not necessary since if we do include the coefficients of z with $Ax = Ax + 0z = b$, the coefficients of z will be all zeros and the only nonzero coefficient of z would always be -1 in the last row. Namely, if we do carry coefficients of z, the corresponding column of coefficients will always be $[0, 0, \cdots, 0, -1]^T$.

At the initial tableau, we append the coefficient d of x to the augmented matrix $[A : b]$ and place the right hand side 0 of (9.19) at the last column. Using elementary row operations we can eliminate the basic variables from the objective function so that we have (9.18) and the first tableau for the linear programming problem, which is called a **simplex tableau**. Namely, we have

$$r_{m+1}x_{m+1} + r_{m+2}x_{m+2} + \cdots + r_nx_n - z = -z_0,$$

from which we check the optimality condition whether $r_j \geq 0$ for every $j = 1, 2, \cdots, n$. Let us use a concrete example to show the implementation of the simplex method, which contains two phases: Phase I shows how to find the initial basic feasible solution; Phase II shows how we begin with the last tableau of Phase I to find the optimal solution of the linear programming problem. We summarize the algorithm after the example.

Example 9.3.1.

$$\begin{array}{ll}
\text{minimize} & 2x + 3y \\
\text{subject to} & x + 2y + z = 8 \\
& 3x + 4y + z = 18 \\
& x,\, y,\, z \geq 0.
\end{array} \qquad (9.20)$$

Solution: To find the initial basic feasible solution, we consider the following auxiliary problem:

$$\begin{array}{ll}
\text{minimize} & u_1 + u_2 \\
\text{subject to} & x + 2y + z + u_1 = 8 \\
& 3x + 4y + z + u_2 = 18 \\
& x,\, y,\, z,\, u_1,\, u_2 \geq 0.
\end{array} \qquad (9.21)$$

The initial simplex tableau is

c_1	c_2	c_3	c_4	c_5	b
1	2	1	1	0	8
3	4	1	0	1	18
0	0	0	1	1	0

where the last row stands for the objective function $u_1 + u_2 - f = 0$. We use elimination to reduce the entries of the last row under the basic variable zero. Then the first simplex tableau is

	c_1	c_2	c_3	c_4	c_5	b
$\xrightarrow{R_3-R_1}$	1	2	1	1	0	8
R_3-R_1	③	4	1	0	1	18
	-4	-6	-2	0	0	-26

$\{c_4, c_5\}$ is the current basis with basic feasible solution $x = (0, 0, 0, 8, 18)$. To have a basic feasible solution, we use two steps to remove $\{c_4, c_5\}$ from the basis. First we bring c_1 into the basis,

$$\begin{aligned}
\epsilon &= \min_i \left\{ \frac{x_i}{a_{i1}} : a_{i1} > 0 \right\} \\
&= \min \left\{ \frac{8}{1}, \frac{18}{3} \right\} \\
&= 6,
\end{aligned}$$

which is achieved at $i_0 = 2$. Note that in this case the basic variables are not in the first columns and the subscript i for x_i indicates the value of the basic variable in the i-th row.

Using the pivot circled in the first simplex tableau, we have

		c_1	c_2	c_3	c_4	c_5	b
$\xrightarrow{\frac{R_2}{3}}$ $\xrightarrow{R_1-R_2}$	R_3-R_2	0	$\boxed{\frac{2}{3}}$	$\frac{2}{3}$	1	$-\frac{1}{3}$	2
		1	$\frac{4}{3}$	$\frac{1}{3}$	0	$\frac{1}{3}$	6
		0	$-\frac{2}{3}$	$-\frac{2}{3}$	0	$\frac{4}{3}$	-2

We have the new basis $\{c_4, c_1\}$. Next we bring c_2 into the basis and obtain

$$\epsilon = \min_i \left\{ \frac{x_i}{a_{i2}} : a_{i2} > 0 \right\}$$
$$= \min \left\{ \frac{2}{\frac{2}{3}}, \frac{6}{\frac{4}{3}} \right\}$$
$$= 3,$$

which is achieved at $i = 1$. We obtain that

		c_1	c_2	c_3	c_4	c_5	b
$\xrightarrow{\frac{3}{2}R_1}$ $\xrightarrow{R_3+\frac{2}{3}R_1}$	$R_2-\frac{4}{3}R_1$	0	1	1	$\frac{3}{2}$	$-\frac{1}{2}$	3
		1	0	-1	-2	1	2
		0	0	0	1	1	0

Therefore $(2, 3, 0, 0, 0)$ is a basic feasible solution such that the auxiliary problem (9.21) is minimized with objective function value 0. Hence $(2, 3, 0)$ is a basic feasible solution of the original problem (9.20).

Notice that the columns c_1, c_2, c_3 and b are common for both (9.20) and (9.21). Therefore we can reuse it for the next steps. We remove the columns for c_4 and c_5 and update the last row by the coefficients of the new objective function.

c_1	c_2	c_3	b
0	1	1	3
1	0	-1	2
2	3	0	0

Eliminating the entries below the basic variables, we have the first simplex tableau:

	c_1	c_2	c_3	b
$\xrightarrow[R_3-2R_2]{R_3-3R_1}$	0	1	①	3
	1	0	-1	2
	0	0	-1	-13

Since the optimality condition is not achieved with $r_3 = -1$, we bring c_3 into the basis:

$$\epsilon = \min_i \left\{ \frac{x_i}{a_{i3}} : a_{i2} > 0 \right\}$$

$$= \min \left\{ \frac{3}{1} \right\}$$

$$= 3,$$

which is achieved at $i = 1$. We have

	c_1	c_2	c_3	b
$\xrightarrow[R_3+R_1]{R_2+R_1}$	0	1	1	3
	1	1	0	5
	0	1	0	-10

which implies that the optimality is achieved and the basic feasible solution is (5, 0, 3) with minimal objective function value 10. □

In summary, we carried out the following algorithm for the linear programming problem (9.9):

Step 1: Write the system $Ax = b$ such that $b \geq 0$, remove any redundant equations such that $\text{rank}(A) = m < n$ and formulate the auxiliary problem (9.10) for a basic feasible solution of (9.9);

Step 2: Write the augmented matrix $[A : I_m : b]$ for $Ax + u = b$ and place the coefficients d for the objective function $d^T x - z = 0$ at the last row with zero in the column for b, ignoring coefficients for z. We call the matrix so obtained the initial tableau for (9.10);

Step 3: Use elementary row operations to reduce the nonzero entries under the columns of the basic variables u into zero. We call the matrix so obtained the first tableau for (9.10);

Step 4: Transfer the extreme points so that u is no longer the basic variable. At each transfer, the column to be moved out and to be replaced by a specified column c_p, $1 \leq p \leq n - m$ is the one corresponding to the basic variable whose value is in row i_0 such that

$$\epsilon = \min_i \left\{ \frac{x_i}{a_{ip}} : a_{ip} > 0 \right\};$$

Step 5: Check the optimality condition, that is, whether coefficients in the last row satisfy $r_j \geq 0$ for $1 \leq j \leq n + m$, and whether the objective function is zero. If the optimal solution is nonzero, (9.9) has no feasible solution. The algorithm stops. Otherwise, the first $n - m$ coordinates of the basic feasible solution currently obtained are a basic feasible solution for (9.9);

Step 6: Delete the columns for u and the last row of the last tableau obtained in Step 5, and update the last row by the coefficients of the new objective function of (9.9). We obtain the initial tableau for (9.9);

Step 7: Use elementary row operations to reduce the nonzero entries under the columns of the basic variables into zero. We call the matrix so obtained the first tableau for (9.9);

Step 8: Check the optimality condition on the current tableau, that is, whether coefficients in the last row satisfy $r_j \geq 0$ for $1 \leq j \leq n$. If yes, then the values of the basic feasible solution are given in the last column and the negative of the objective function value $-z_0$ is in the right lower corner;

Otherwise, transfer the extreme points to achieve optimality. At each transfer, the column to be moved out and replaced by a specified column c_p, $1 \leq p \leq n$ is the row number i_0 such that

$$\epsilon = \min_i \left\{ \frac{x_i}{a_{ip}} : a_{ip} > 0 \right\}.$$

Exercise 9.3.2.

1. Determine whether the following linear program has a basic feasible solution:

$$
\begin{array}{llrcl}
\text{minimize} & & 2x + 3y & & \\
\text{subject to} & & x + 2y + z & = & 8 \\
& & 3x + 4y + z & = & 28 \\
& & x, y, z & \geq & 0.
\end{array}
$$

2. Solve the following linear programming problem.

$$\begin{array}{lrcl} \text{minimize} & x + y & & \\ \text{subject to} & x + 2y + z & = & 8 \\ & 3x + 4y + z & = & 18 \\ & x,\, y,\, z & \geq & 0. \end{array}$$

3. Solve the following linear programming problem.

$$\begin{array}{lrcl} \text{minimize} & 2x + 3y & & \\ \text{subject to} & x + 2y & \geq & 8 \\ & 3x + 4y & \geq & 18 \\ & x,\, y & \geq & 0. \end{array}$$

4. Solve the following linear programming problem.

$$\begin{array}{lrcl} \text{maximize} & 2x + 3y & & \\ \text{subject to} & x + 2y & \leq & 8 \\ & 3x + 4y & \leq & 18 \\ & x,\, y & \geq & 0. \end{array}$$

5. Solve the following linear programming problem.

$$\begin{array}{lrcl} \text{minimize} & 2x - 3y & & \\ \text{subject to} & x + 2y + z & = & 8 \\ & 3x + 4y + z & = & 18 \\ & x,\, y,\, z & \geq & 0. \end{array}$$

6. Solve the following linear programming problem.

$$\begin{array}{lrcl} \text{maximize} & -2x + 3y & & \\ \text{subject to} & x + 2y + z & = & 8 \\ & 3x + 4y + z & = & 18 \\ & x,\, y,\, z & \geq & 0. \end{array}$$

Index